DATE DUE

THE
Physics Quick
Reference Guide

THE
PHYSICS QUICK
REFERENCE GUIDE

E. Richard Cohen
Rockwell International Corporation
Thousand Oaks, California

AIP PRESS

American Institute of Physics   **Woodbury, New York**

AIP Press
American Institute of Physics
500 Sunnyside Boulevard
Woodbury, NY 11797-2999

Library of Congress Cataloging-in-Publication Data
Cohen, E. Richard, 1922–
    The physics quick reference guide / E. Richard Cohen.
        p. cm.
    Includes bibliographical references and index.
    ISBN 1-56396-143-1
    1. Physics--Handbooks, manuals, etc. I. Title.
QC61.C65 1995   95-24953
530--dc20   CIP

10 9 8 7 6 5 4 3 2 1

CONTENTS

CONTENTS

PREFACE

As one of the events to mark its 50th anniversary in 1981, the American Institute of Physics assembled a handbook, *Physics Vade Mecum* [revised and reissued as *A Physicist's Desk Reference* (PDR) in 1986], consisting of 21 chapters covering a wide range of topics from acoustics to thermophysics, plus an introductory general section.

This volume is an expansion of that first chapter and is intended to serve as a compendium of useful material applicable to the broad range of physics. By its nature this implies that a major portion of the material is mathematical. This is partially a reflection of personal bias, but I hope that it is more nearly a reflection of the generality and pervasiveness of the topics. I have also specifically removed some material in Chapter 1 of the PDR that I felt was out of date and that reflected an earlier stage of physics characterized by "love and string and sealing wax," when experimental physics involved building one's own equipment, blowing one's own glassware, and scrounging army and navy surplus for bombsights and radar systems. The discussion of physical quantities and their symbols and units has been expanded. The International System of units (SI) is used exclusively in all tables. Some material has been included here from the other chapters of the PDR for completeness.

This little handbook cannot, nor is it intended to, substitute for more complete collections such as *Handbook of Physics*, E. U. Condon and H. Odishaw; *American Institute of Physics Handbook*, D. E. Gray; *Handbook of Chemistry and Physics*, CRC Press; *Handbook of Mathematical Functions*, M. Abramowitz and I. G. Stegun; *Tables of Integrals, Series, and Products*, I. S. Gradsteyn and I. M. Ryzhik; or several other compilations of mathematical and physical data. Numerical tables of mathematical functions have been purposely omitted since they have, in general, been made obsolete by the pocket calculator.

The tables in Chapter 7 are intended to present only representative data rather than complete coverage. The tables of Fourier and Laplace transforms in Chapter 8 give only the most basic relationships but with hints as to how they may be extended. Obviously these chapters are not a replacement for more extensive sources.

ix

I wish to thank all those who have helped with useful data or comments on the various drafts; although some of these were anonymous, I can name in particular (without intending to slight all those others who also contributed): S. C. Abrahams, H. Barschall, R. T. Beyer, G. C. Carter, I. Goldberg, and R. B. Goldfarb.

1

PHYSICAL QUANTITIES, SYMBOLS, AND UNITS

1.1 Numbers

1.1.1 Writing numbers

Although both American and British English use a dot (.) as the decimal point, the comma (,) is used in most European languages, including those using the Cyrillic alphabet. The International Organization for Standardization (ISO) prefers the comma, and uses it even in material published in English.[1] Since the reverse convention with regard to the comma or the dot as a spacer in writing long numbers also divides English from most of the rest of the world, to avoid possible misinterpretation, no symbol other than the decimal point (either dot or comma) should be used to punctuate numbers. The centered dot (·), which has sometimes been used in the past in British English, should never be used as a decimal point in scientific writing.

Decimal points should never "go naked." There should always be at least one digit both before and after the decimal point. An integer should not be terminated by a decimal point and if the magnitude of the number is less than unity the decimal point should be preceded by a zero.

137 or 137.0   but not   137.;   0.036   not   .036

To facilitate the reading of long numbers (greater than four digits either to the right or to the left of the decimal point) the digits may be separated into groups (usually of three, but sometimes —particularly in tables— four or five, digits) by thin spaces. Instead of a single final digit, the last group may be increased.

299 792 458;   1.234 567 8   or   1.234 5678

Dates should always be written without a space (e.g., 3761 B.C., 1995).

1.1.2 Arithmetical operations

The sign for multiplication of numbers is a cross ($\times$); the multiplication sign may be omitted when the numbers are enclosed in parentheses or braces.

Examples:
$$1.2 \times 3.45 \quad \text{or} \quad (137.036)(273.16).$$

Division of one number by another number may be indicated either by a horizontal bar or by a solidus (/), or by writing it as the product of the numerator and the inverse first power of the denominator. In such cases the number under the inverse power should always be placed in parentheses, brackets, or other sign of aggregation.

Examples: $\qquad \dfrac{137}{273.16}, \qquad 137/273.16, \qquad 137\,(273.16)^{-1}.$

When the solidus is used, brackets or parentheses should be included if necessary to clarify where the numerator starts or where the denominator ends.

1.2 Physical quantities

A physical quantityan attribute of a phenomenon, body or substance that may be distinguished qualitatively and determined quantitatively.[2]

Thus a physical quantity is an attribute that can be measured (actually or conceptually) and to which a numerical value can be assigned.

The term *physical quantity* is used with two different but related connotations. One refers to the general concept, the other to a specific example of that concept. The question, "What is the diameter of this sphere?" can be answered by a general statement, "The maximum chord between two points on its surface," or by a general statement, "12.3 mm." In the absence of a vocabulary that distinguishes between the two concepts, the context must provide the meaning.

The value of a physical quantity is expressed as the product of a numerical value and a unit of measurement:

$$X = \text{physical quantity} = \text{number} \times \text{unit} = \{X\}\,[X],$$

where $\{X\}$ denotes the numerical value of X and $[X]$ denotes the unit of X. Neither the name nor the symbol for a physical quantity should imply any particular choice of unit. Thus, as a specific example, in spite of historical precedent the term "voltage" should be avoided in favor of the terms "potential difference" or "electrical potential."

1.2.1 Symbols for quantities

Symbols for physical quantities are usually single letters of the Latin, Greek, or other alphabet in *sloping* or *italic* type that may be modified by the use of subscripts and superscripts. In general, subscripts and superscripts referring to physical quantities are printed in sloping type while descriptive or numerical subscripts and superscripts are printed in upright type.

Examples:
C_p for the heat capacity at constant pressure, but C_g for the heat capacity of a sample in the gaseous state

$$\sum_r a_r x^r, \qquad \text{but } a_0 + a_1 x + a_2 x^2 + \ldots,$$

a_r where the subscript r indicates "relative."

Complicated subscripts or superscripts may be avoided by using parentheses or brackets: instead of $\rho_{\mathrm{H_2SO_4,20\,°C}}$ write $\rho(\mathrm{H_2SO_4}, 20\,°\mathrm{C})$.

It is usual to use standard weight type symbols for scalars, and boldface type symbols for aggregates (vectors and tensors). A distinction is made between boldface type for the vector as an entity and standard weight type for the components of that vector: the vector $\boldsymbol{A}$ has components A_k. IUPAP recommends that vectors representing physical quantities be printed in slanted type, $\boldsymbol{a}$, $\boldsymbol{A}$, while the *AIP Style Guide* sets all vectors as upright boldface, $\mathbf{a}$, $\mathbf{A}$. Tensors can be printed in bold sansserif type, e.g., T.

When distinctive type is not available, a vector may be indicated by an arrow above the symbol, $\vec{a}$, $\vec{B}$, and a tensor by a double arrow $\vec{\vec{S}}$, or by a double headed arrow $\overset{\leftrightarrow}{S}$. Extending this pattern to higher-dimensional tensors obviously becomes awkward; in such cases an index notation should be used uniformly for vectors and tensors:

$$A_i, \qquad S_{ij}, \qquad R^{ij}_{kl}, \qquad R^{i}_{.\,jk\,.}{}^{l}.$$

1.2.2 Quantity algebra

Physical quantities may be combined by addition or subtraction only when they have the same dimensions, but they may combine by multiplication or division following the usual rules of algebra, which then apply separately and equally to the numerical values and to the units.

Addition and Subtraction of Two Physical Quantities: $a + b$ and $a - b$;

Multiplication of Two Physical Quantities: $ab, a \cdot b, a \times b$;

Division of One Quantity by Another Quantity: $\dfrac{a}{b}, a/b, ab^{-1}$, or any other way of writing the product of a and b^{-1}.

These procedures can be extended to cases where one of the quantities, or both, are themselves products, quotients, sums, or differences of other quantities, using parentheses and brackets as necessary in accordance with the rules of mathematics. When a solidus is used to separate the numerator from the denominator, brackets should be inserted if there is any doubt where the numerator starts or where the denominator ends.

Examples:

Expression with a horizontal bar	Same expression with a solidus
$\dfrac{a}{bcd}$	a/bcd or $a/(bcd)$
$\dfrac{2}{9}\sin kx$	$(2/9)\sin kx$
$\dfrac{a}{b} - c$	$a/b - c$
$\dfrac{a}{b-c}$	$a/(b-c)$
$\dfrac{a+b}{c-d}$	$(a+b)/(c-d)$
$\dfrac{a}{b} + \dfrac{c}{d}$	$a/b + c/d$ or $(a/b) + (c/d)$

Parentheses, braces, or brackets should be used in order to define unambiguously the argument of a mathematical function:

Examples:

$$\sin\{2\pi(x - x_0)/\lambda\}, \qquad \exp\{(r - r_0)/\sigma\},$$
$$\exp[-V(r)/kT], \qquad \sqrt{(G/\rho)}.$$

Parentheses may be omitted when the argument is a single quantity or a simple product: e.g., $\sin\theta$, $\tan kx$.

A horizontal over bar used with the square root sign defines the outermost level of aggregation: $\sqrt{(ab)}$ or $\sqrt{ab}$, but for a compound expression the form (for example) $\sqrt{[g(t) + h(t)]/H(t)}$ may be preferable to $\sqrt{\{[g(t) + h(t)]/H(t)\}}$.

1.2.3   Units and dimensions

The *value of the physical quantity* X is independent of the unit used to express it, but the numerical value $\{X\}$ will depend on the unit. If the unit is changed from $[X]_1$ to $[X]_2 = a[X]_1$, the numerical value changes from $\{X\}_1$ to $\{X\}_2 = a^{-1}\{X\}_1$.

$$X = \{X\}_1[X]_1 = \{X\}_2[X]_2 = \{X\}[X].$$

Example: $\lambda = 6.058{\times}10^{-7}\,\mathrm{m} = 605.8\,\mathrm{nm} = 23.8504\,\mu\mathrm{in},$

which may also be expressed by writing $\{\lambda\}_\mathrm{m} = 6.058{\times}10^{-7}$, $\{\lambda\}_\mathrm{nm} = 605.8$, $\{\lambda\}_{\mu\mathrm{in}} = 23.8504$.

The *dimension* of a physical quantity is the set of all units by which the physical quantity can be expressed. Thus the dimension of length, L, is the set [m, mm, km, in, ft, yd, Å, ly, ...]; the dimension of energy, E, is the set [erg, joule, kilowatt-hour, BTU, calorie, ...].

The symbol $[X]$, as defined by Maxwell, had both a specific and a general connotation; in a specific sense, it represents the unit of measurement (in the example above, $[\lambda]_1 = \mathrm{m}$, $[\lambda]_2 = \mathrm{nm}$, $[\lambda]_3 = \mu\mathrm{in}$), in the general sense it signified the dimension of X, dim X (in the example above, dim $\lambda = \mathrm{L}$).

When a physical quantity is divided by its unit the result is a pure number, the numerical value of the quantity in the specified unit system: $\{X\} = X/[X]$. For the wavelength in the example above, one has

$$\lambda/\mathrm{m} = 6.058{\times}10^{-7}, \qquad \lambda/\mathrm{nm} = 605.8.$$

Such a form (quantity/unit) is useful in labeling the axes of a graph or the columns in a table to give an unambiguous indication of the meaning of the associated numbers.

A *conversion factor* f changes a quantity from one unit basis to another; it is an expression whose physical value is unity. Thus, from the expressions above one obtains:

$$f = \frac{6.058{\times}10^{-7}\,\mathrm{m}}{605.8\,\mathrm{nm}} = 10^{-9}\,\mathrm{m/nm} = 1.$$

Since a conversion factor has the value 1, any physical quantity may be multiplied by conversion factors at will.

1.2.4   Symbols for units

The name of a unit is a common noun and is not capitalized, even if it is derived from a proper name. When the name of the unit is derived from a proper name its symbol is composed of one or two letters, the first of which is capitalized. If the name is not derived from a proper name the symbol is in lower case. (An exception to this rule is the allowed symbol

L, in addition to l, for liter because of the possibility of confusing l with the number 1.)

Examples: meter, m   mole, mol   ampere, A   weber, Wb.

Symbols for units do not contain a period, and remain unaltered in the plural.

Examples: 1 m   2 m   3.1416 m   *and not*   2 m. or 2 ms.

The multiplication of units is written, as recommended by the International Organization for Standardization[1] (ISO) and accepted by both IUPAP[3] and IUPAC[4] as well as the *AIP Style Guide*[5], in either of the following two forms, e.g.:

$$N \cdot m \qquad N\, m$$

while the American National Standard[6] ANSI/IEEE Std 268-1982 states that in US practice the first form (the centered dot) is to be preferred.

1.3   Systems of units

In a physical system consisting of a set of quantities and the relational equations connecting them, an arbitrary number of quantities are defined by convention to be dimensionally independent. These quantities form the set of *base quantities* for the system. All other physical quantities are *derived quantities*, defined in terms of the base quantities and expressed algebraically as products of powers of the base quantities. In the field of mechanics, three base quantities (length L, mass M, and time T) are usually considered to be adequate, while heat and thermodynamics introduces temperature, dimension Θ. The dimension of any other quantity is then expressed in terms of the dimensional product of the base dimensions:

$$\dim Q = L^{\alpha} M^{\beta} T^{\gamma} \Theta^{\delta}.$$

The powers to which the various base quantities or base dimensions are raised are called the *dimensional exponents*.

A quantity that arises from dividing one physical quantity by another with the same dimension (e.g., relative density, refractive index), or, more generally, any quantity that has a dimension whose dimensional exponents are all equal to zero, $L^0 M^0 T^0 \Theta^0$, has a dimension symbolized by the number 1. Such a quantity, a quantity of dimension one, is often called a *dimensionless quantity*.

The mathematical relation among physical quantities may be a simple monomial or a sum of such terms, but each term of the sum must have the same dimension. Such a relation can always be expressed as $\phi(x_1, x_2, \dots) = 0$ in which the arguments x_j are variables of dimension one. When a physical quantity Y is expressed as a function of other variables, the relation must then be expressible (although such a form is usually not explicitly shown) as $Y = AF(x_1, x_2, \dots)$ where F and its arguments x_j are quantities of dimension one and A has the same dimension as Y.

A set of conventionally defined samples of the base quantities are chosen as *base units* to form the foundation for a *system of units*. While it is possible to define an independent unit U_X for every quantity, it is much more convenient to relate the units of the derived quantities to the base units as products of powers, corresponding to the expressions in the system of quantities:

$$U_Q = k_Q U_L^\alpha U_M^\beta U_T^\gamma U_\Theta^\delta.$$

When all derived units are expressed in terms of the base units by relations with numerical factors k_Q equal to unity, the system is said to form a *coherent set* of units; in such a system, the numerical equations and the physical equations have the same form. Since there are no numerical quantities introduced in the definition of derived units, the equation expressing the value of a physical quantity has the same form as the equation expressing the numerical value of the quantities.

Derived units and their symbols are expressed algebraically in terms of base units by means of the mathematical signs for multiplication and division. Some derived units receive special names and symbols, which can then be used in forming names and symbols of other derived units. A derived unit can be expressed in different ways using the names of base units and the special names of derived units. A given physical quantity has a unique dimension in terms of base dimensions, even though its unit may be expressed in more than one way using appropriate combinations of base and derived units. A given unit or dimension, however, may correspond to more than one quantity; e.g., the quantities *kinematic viscosity* (η/ρ) and *diffusion coefficient* both have the dimension $L^2 T^{-1}$, (*length squared divided by time*), while *heat capacity* and *entropy* both have the dimension $ML^2 T^{-2} \Theta^{-1}$, (*energy per degree*).

1.4 Dimensional and "dimensionless" ratios

1.4.1 Coefficients and factors

When a quantity A is proportional to another quantity B, the relation is expressed by an equation of the form $A = kB$. The quantity k is called a "factor" or "index" if A and B have the same dimension and k has dimension one, and a "constant," "coefficient," or "modulus" if A and B have different dimensions.

Examples:

$$E = A_H(B \times J), \qquad A_H : \quad \text{Hall coefficient,}$$
$$\sigma = E\epsilon, \qquad E : \quad \text{Young's modulus,}$$
$$J = -D\,\nabla n, \qquad D : \quad \text{diffusion coefficient,}$$
$$L_{12} = k\sqrt{L_1 L_2}, \qquad k : \quad \text{coupling factor,}$$
$$F = \mu F_{\mathrm{n}}, \qquad \mu : \quad \text{friction factor .}$$

1.4.2   Parameters, numbers, and ratios

A combination of physical quantities used to characterize the behavior or properties of a physical system is called a *parameter*; if, however, the quantity has dimension one it is often referred to as a *number* or a *ratio*. If such a ratio is inherently non-negative and less than or equal to 1 it is often called a *fraction*.

Examples:

$$\text{Grüneisen parameter :} \qquad \gamma = \alpha/\kappa\rho c_V,$$
$$\text{Reynolds' number :} \qquad Re = \rho v l/\eta,$$
$$\text{Mobility ratio :} \qquad b = \mu_-/\mu_+,$$
$$\text{Molar fraction :} \qquad x_B = n_B \Big/ \sum_j n_{B_j}.$$

1.5   Symbols and nomenclature

1.5.1   Chemical elements

Symbols for chemical elements are written in upright type; the symbol is not followed by a period.

Examples: americium, Am;   iodine, I;   phosphorus, P.

The nucleon number (mass number, baryon number) of a nuclide is shown as a left superscript (e.g., ^{14}N). In nuclear physics, when there will be no confusion with molecular compounds, a left subscript is sometimes used to indicate the number of protons and a right subscript to indicate the number of neutrons in the nucleus; although these subscripts are redundant they are often useful (e.g., $^{235}_{92}$U$_{143}$). Since the subscripts are redundant, they are usually omitted; the right subscript should not be included unless the left subscript is also present.

The right subscript position is used to indicate the number of atoms of a nuclide in a molecule.

Examples: C_2H_6, $\quad C_2{}^1H_4{}^2H_2$.

The right superscript position can be used, if required, to indicate a state of ionization (e.g., Ca_2^+, PO_4^{3-}), an excited *atomic* state (e.g., He^*), or a metastable *nuclear* state (e.g., $^{118}Ag^m$). In nuclear physics, the metastable state often is treated as a distinct nuclide: e.g., ^{118m}Ag.

Roman numerals are used in two different ways:

(i) The spectrum of a z-fold ionized atom is specified by the small capital Roman numeral corresponding to $z+1$, written on the line with a thin space following the chemical symbol.

Examples: O I (neutral oxygen), $\quad$ Ca II, $\quad$ O III .

(ii) Roman numerals in the right superscript position are used to indicate the oxidation number.

Examples: $Pb_2^{II}Pb^{IV}O_4$, $\quad K_6Mn^{IV}Mo_9O_{32}$.

1.5.2 Nuclear particles

Table 1.1 Symbols for nuclear particles. The common designations for particles used as projectiles or products in nuclear reactions are listed. In addition to the symbols given, an accepted designation for a general heavy ion (where there is no chance of ambiguity) is HI.

Photon	γ	Nucleon	N
Neutrino	$\nu, \nu_e, \nu_\mu, \nu_\tau$	Neutron	n
Electron	e, β	Proton ($^1H^+$)	p
Muon	μ	Deuteron ($^2H^+$)	d
Tauon	τ	Triton ($^3H^+$)	t
Pion	π	Helion ($^3He^{2+}$)	h [a]
		Alpha particle ($^4He^{2+}$)	α

[a] The symbol τ has been used in older literature for the helion, but that symbol should be reserved for the tauon (heavy lepton).

The charge of a particle may be indicated by adding a superscript $^+, ^0, ^-$ to the symbol for the particle.

Examples: π^+, π^0, π^-; $\quad e^+, e^-$; $\quad \beta^+, \beta^-$.

If no charge is indicated in connection with the symbols p and e, these symbols refer to the positive proton and the negative electron, respectively. The bar ⁻ or the tilde ˜ above the symbol for a particle is used to indicate the corresponding anti-particle; the notation $\bar{p}$ is preferable to p⁻ for the anti-proton, but both $\bar{e}$ and e⁺ (or $\bar{\beta}$ and β^+) are commonly used for the positron.

1.5.3 Fundamental particles

Although IUPAP recommends that the symbols for particles be printed in upright type [so that, for example, the symbol e (upright) for the electron will not be confused with the symbol e (slanted) for the elementary charge], the Particle Data Group[7] and the *AIP Style Guide*[5] use slanted type for particle symbols.

The names of many fundamental particles are simply the names for their symbols. The names "up," "down," "charm," "strange," "top (truth)" and "bottom (beauty)" for quarks are to be considered only as mnemonics; the names of quarks are the symbols themselves.

Table 1.2 Symbols for fundamental particles. Except for proton p, neutron n, and electron e, the symbols for the fundamental particles in this table follow the recommendations of the Particle Data Group (Ref. 7).

Gauge bosons	γ, g, $W^\pm$, Z
Leptons	$e^\pm$, $\nu_e^\pm$, $\mu^\pm$, $\nu_\mu^\pm$, $\tau^\pm$, $\nu_\tau^\pm$
Quarks (q)	u, d, c, s, b, t
Mesons ($q\bar{q}$)	
unflavored ($S = C = B = 0$)	$\pi^\pm$, π^0, η, ρ, ω, η', f, a,
	ϕ, h, b, f'
strange ($S = \pm1$, $C = B = 0$)	$K^\pm$, K^0 (K_S, K_L)
charmed ($S = 0, C = \pm1$)	$D^\pm$, D^0
charmed strange ($C = S = \pm1$)	$D_s^\pm$
bottom ($S = 0, B = \pm1$)	$B^\pm$, B^0
$c\bar{c}$ mesons	η_c, J/ψ, χ_c
$b\bar{b}$ mesons	Υ, χ_b
Baryons (qqq)	
($S = 0$)	p, n, N^+, N^0, Δ
($S = -1$)	Λ^0, $\Sigma^\pm$, Σ^0
($S = -2$)	Ξ^0, Ξ^-
($S = -3$)	Ω^-
charmed baryons ($C = +1$)	Λ_c^+, Σ_c, Ξ_c^+, Ξ_c^0, Ω
bottom baryon ($B = -1$)	Λ_b^0

1.5.4   Spectroscopic notation

A letter symbol indicating a quantum number of a *single particle* is printed in lowercase upright type. A letter symbol indicating a quantum number of a *system* is printed in capital upright type.

Atomic spectroscopy: The letter symbols indicating the orbital angular momentum quantum number are

$l =$	0	1	2	3	4	5	6	7	8	9	10	11	...
symbol:	s	p	d	f	g	h	i	k	l	m	n	o	...

$L =$	0	1	2	3	4	5	6	7	8	9	10	11	...
symbol:	S	P	D	F	G	H	I	K	L	M	N	O	...

A right subscript attached to the angular momentum symbol indicates the total angular momentum quantum number j or J. A left superscript indicates the spin multiplicity, $2s + 1$ or $2S + 1$.

Examples:

$$\text{d}_{\frac{3}{2}} - \text{electron} \qquad (j = \tfrac{3}{2}),$$
$$^3\text{D} - \text{term} \qquad (\text{spin multiplicity} = 3),$$
$$^3\text{D}_2 - \text{level} \qquad J = 2.$$

An atomic electron configuration is indicated symbolically by:

$$(nl)^k (n'l')^{k'} (n''l'')(n''l'')^{k''} \ \dots$$

in which $k, k', k'', \dots$ are the numbers of electrons with principal quantum numbers $n, n', n'', \dots$ and orbital angular momentum quantum numbers $l, l', l'', \dots$, respectively. In a specific instance, instead of the orbital angular momentum quantum number l, one uses the quantum number symbols s ($l = 0$), p ($l = 1$), d ($l = 2$), f ($l = 3$), ..., and the parentheses are usually omitted.

Example: $1\text{s}^2 2\text{s}^2 2\text{p}^3$.

An atomic state is specified by giving all of its quantum numbers. An atomic *term* is specified by L and S and an atomic *level* by L, S, and J. An atomic *state* is specified by L, S, J, and M_J, or by L, S, M_S, and M_L.

Molecular spectroscopy: For *linear molecules* the letter symbols indicating the quantum number of the component of electronic orbital angular momentum along the molecular axis are

$$\begin{aligned} \lambda = \quad & 0 \quad 1 \quad 2 \quad \ldots \\ \text{symbol:} \quad & \sigma \quad \pi \quad \delta \quad \ldots \\[4pt] \Lambda = \quad & 0 \quad 1 \quad 2 \quad \ldots \\ \text{symbol:} \quad & \Sigma \quad \Pi \quad \Delta \quad \ldots \end{aligned}$$

A left superscript indicates the spin multiplicity. For molecules having a symmetry center, the parity symbol g (*gerade*) or u (*ungerade*) indicating, respectively, symmetric or antisymmetric behavior on inversion is attached as a right subscript. A $^+$ or $^-$ sign attached as a right superscript indicates the symmetry with regard to reflection in any plane through the symmetry axis of the molecule (e.g., Σ_{g}^+, Π_{u}).

The letter symbols indicating the quantum number of vibrational angular momentum are

$$\begin{aligned} l = \quad & 0 \quad 1 \quad 2 \quad 3 \quad \ldots \\ \text{symbol:} \quad & \Sigma \quad \Pi \quad \Delta \quad \Phi \quad \ldots \end{aligned}$$

Nuclear spectroscopy: The spin and parity assignment of a nuclear state is J^π where the parity symbol π is $+$ for even parity and $-$ for odd parity (e.g., 3^+, 2^-).

A shell model configuration is indicated symbolically by

$$\nu(nl_j)^\kappa (n'l'_{j'})^{\kappa'} \ldots \pi(n''l''_{j''})^{\kappa''} (n'''l'''_{j'''})^{\kappa'''} \ldots$$

where the letter π refers to the proton shell and the letter ν to the neutron shell. Negative values of a superscript indicate holes in a completed shell. Instead of $l = 0, 1, 2, 3, \ldots$ one uses the symbols s, p, d, f, $\ldots$ as in atomic spectroscopy (except for $l = 7$ which is denoted by k in atoms and by j in nuclei).

Example: the nuclear configuration: $\nu(2\mathrm{d}_{\frac{5}{2}})^6 \, \pi(2\mathrm{p}_{\frac{1}{2}})^2 (1\mathrm{g}_{\frac{9}{2}})^3$.

When the neutrons and protons are in the same shell with well-defined isospin T, the notation $(nl_j)^\alpha$ is used where α denotes the total number of nucleons [e.g., $(1\mathrm{f}_{\frac{7}{2}})^5$].

Spectroscopic transitions: The upper (higher energy) level and the lower (lower energy) level of a transition are indicated, respectively, by $'$ and $''$.

Examples: $h\nu = E' - E''$, $\sigma = T' - T''$.

Unfortunately, the designation of spectroscopic transitions is not uniform. In *atomic* spectroscopy[8] the convention is to write the *lower* state first and the *upper* state second; in *molecular* and *polyatomic* spectros-copy[9] the convention is reversed, and one writes the *upper* state first and the *lower* state second.

In either case the two state designations are connected by a dash — or, if it is necessary to distinguish between an absorbing and an emitting transition, by arrows $\leftarrow$ and $\rightarrow$. If there is a chance of ambiguity, the convention being used with regard to the ordering of the states should be clearly stated.

Examples:

$$2\,^2S_{\frac{1}{2}} - 4\,^2P_{\frac{3}{2}} \qquad \text{atomic transition ,}$$
$$(J', K') \leftarrow (J'', K'') \qquad \text{molecular rotational absorption .}$$

The difference between two quantum numbers is that of the upper state minus that of the lower state.

Example: $\qquad \Delta J = J' - J''.$

The branches of the rotation–vibration band are designated as

branch		O	P	Q	R	S
$\Delta J = J' - J''$		-2	-1	0	$+1$	$+2$.

1.5.5 Nomenclature conventions in nuclear physics

Nuclide: A species of atoms identical as regards atomic number (pro-ton number) and mass number (nucleon number) should be indicated by the word "nuclide," not "isotope" which implies a relationship like "sis-ter." Different nuclides having the same atomic number (proton number) are called *isotopic nuclides* or *isotopes*. (Different nuclides with the same neutron number have sometimes been designated as "isotones.") Different nuclides having the same mass number are *isobaric nuclides* or *isobars*.

The symbolic expression representing a nuclear reaction follows the pat-tern:

$$\text{initial} \atop \text{nuclide} \left({\text{incoming particle} \atop \text{or photon}} , {\text{outgoing particle(s)} \atop \text{or photon(s)}} \right) {\text{final} \atop \text{nuclide.}}$$

Examples:

$$^{14}N\,(\alpha, p)\,^{17}O, \qquad ^{59}Co\,(n, \gamma)\,^{60}Co,$$
$$^{23}Na\,(\gamma, 3n)\,^{20}Na, \qquad ^{31}P\,(\gamma, pn)\,^{29}Si.$$

Characterization of interactions:

Multipolarity of a transition:

electric or magnetic monopole	E0 or M0 ,
electric or magnetic dipole	E1 or M1 ,
electric or magnetic quadrupole	E2 or M2 ,
electric or magnetic octopole	E3 or M3 ,
electric or magnetic 2^n-pole	En or Mn .

Designation of parity change in a transition:

transition *with* parity change :	(yes) ,
transition *without* parity change :	(no) .

Notation for covariant character of coupling:

S	Scalar coupling,	A	Axial vector coupling,
V	Vector coupling,	P	Pseudoscalar coupling.
T	Tensor coupling,		

Polarization conventions:

Sign of polarization vector (Basel convention): In a nuclear interaction the positive polarization direction for particles with spin $\frac{1}{2}$ is taken in the direction of the vector product

$$\boldsymbol{k}_{\mathrm{i}} \times \boldsymbol{k}_{\mathrm{o}}$$

where $\boldsymbol{k}_{\mathrm{i}}$ and $\boldsymbol{k}_{\mathrm{o}}$ are the wave vectors of the incoming and outgoing particles, respectively.

Description of polarization effects (Madison convention): In the symbolic expression for a nuclear reaction A(b,c)D, an arrow placed over a symbol denotes a particle which is initially in a polarized state or whose state of polarization is measured.

Examples:

A($\vec{\mathrm{b}}$, c)D	polarized incident beam,
A($\vec{\mathrm{b}}$,$\vec{\mathrm{c}}$)D	polarized incident beam; polarization of the outgoing particle c is measured (polarization transfer),
A(b,$\vec{\mathrm{c}}$)D	unpolarized incident beam; polarization of the outgoing particle c is measured,
$\vec{\mathrm{A}}$(b, c)D	unpolarized beam incident on a polarized target,
$\vec{\mathrm{A}}$(b,$\vec{\mathrm{c}}$)D	unpolarized beam incident on a polarized target; polarization of the outgoing particle c is measured,
A($\vec{\mathrm{b}}$,c)$\vec{\mathrm{D}}$	polarized incident beam; polarization of the residual nucleus is measured.

References

[1]   *ISO Standards Handbook 2*, 3rd ed., International Organization for Standardization, (Geneva 1993). [This Handbook combines ISO Standard 1000 (Units) and the series on standard symbols in several fields of physics (ISO Standards 31.0 – 31.12).]

[2]   *International Vocabulary of Basic and General Terms in Metrology*, (Jointly sponsored by ISO, IEC, BIPM, OIML, IUPAP, IUPAC, IFCC), Geneva 1993.

[3]   E. R. Cohen and P. Giacomo, Symbols, Units, Nomenclature and Fundamental Constants in Physics, [*Physica*, **146A**, (1987)] (IUPAP-25).

[4]   *Quantities, Units and Symbols in Physical Chemistry* edited by I. M. Mills and T. Cvitaš (Blackwell Scientific, Oxford 1993).

[5]   *AIP Style Guide*, (American Institute of Physics, New York, 1990).

[6]   ANSI/IEEE, *American National Standard for Metric Practice*, (Std 268-1992), IEEE Standards Coordinating Committee 14 (IEEE, New York, 1992).

[7]   Particle Data Group, Review of Particle Properties, [Phys. Rev. D **50**, 1173 (1994)].

[8]   R. D. Cowan, *The Theory of Atomic Structure and Spectra* (University of California Press, 1981).

[9]   Report on Notation for the Spectra of Polyatomic Molecules, [J. Chem. Phys. **23**, 1997 (1955)].

2

THE INTERNATIONAL SYSTEM OF UNITS (SI)

The International System of Units (Système International d'Unités, with the international abbreviation SI), is the modern version of the metric system adopted by the Conférence Générale des Poids et Mesures (General Conference of Weights and Measures, CGPM) in 1960. It is a coherent system with seven dimensionally independent base units. These units and the dates of the adoption of their present definitions by CGPM are

meter (metre): The meter is the length of the path traveled by light in vacuum during a time interval of $1/299\,792\,458$ of a second. [17th CGPM (1983), Resolution 1.]

kilogram: The kilogram is the unit of mass; it is equal to the mass of the international prototype of the kilogram. [1st CGPM (1889) and 3rd CGPM (1901).]

second: The second is the duration of $9\,192\,631\,770$ periods of the radiation corresponding to the transition between the two hyperfine levels of the ground state of the cesium-133 atom. [13th CGPM (1967), Resolution 1.]

ampere: The ampere is that constant current which, if maintained in two straight parallel conductors of infinite length, of negligible circular cross section, and placed 1 meter apart in vacuum, would produce between these conductors a force equal to 2×10^{-7} newton per meter of length. [9th CGPM (1948), Resolutions 2 and 7.]

kelvin: The kelvin, the unit of thermodynamic temperature, is the fraction $1/273.16$ of the thermodynamic temperature of the triple point of water. [13th CGPM (1967), Resolution 4.]

The 13th CGPM (1967, Resolution 3) also decided that the unit kelvin and its symbol K should be used to express both the thermodynamic temperature and an interval or a difference of temperature.

Table 2.1 SI base units.

	SI unit	
Base quantity	Name	Symbol
length	meter	m
mass	kilogram	kg
time	second	s
electric current	ampere	A
thermodynamic temperature	kelvin	K
amount of substance	mole	mol
luminous intensity	candela	cd

In addition to the thermodynamic temperature (symbol T) there is also the Celsius temperature (symbol t) defined by the equation

$$t = T - T_\circ,$$

where $T_\circ = 273.15\,\mathrm{K}$. Celsius temperature is expressed in degree Celsius (symbol, $^\circ$C). The unit degree Celsius is equal to the unit kelvin, and a temperature interval or a difference of temperature may be expressed in kelvins as well as in degrees Celsius.

By definition, the triple point of water is $t_{tp} = 0.01\,^\circ$C; the freezing point of water is then $T_f = 0\,^\circ$C with an uncertainty of the order of $10^{-4}\,$K, and the boiling point (at 1 atmosphere pressure) is $T_b = 99.974\,^\circ$C (uncertainty of the order of $10^{-3}\,$K). The Celsius scale is therefore *not* a "centigrade" scale.

mole: 1.   The mole is the amount of substance of a system which contains as many elementary entities as there are atoms in 0.012 kilogram of carbon 12.

2.   When the mole is used, the elementary entities must be specified and may be atoms, molecules, ions, electrons, other particles, or specified groups of such particles. [14th CGPM (1971), Resolution 3.]

In this definition, it is understood that the carbon 12 atoms are unbound, at rest, and in their ground state.

candela: The candela is the luminous intensity, in a given direction, of a source that emits monochromatic radiation of frequency 540×10^{12} hertz and that has a radiant intensity in that direction of 1/683 watt per steradian [16th CGPM (1979), Resolution 3].

Table 2.2 Prefixes used in SI.   These prefixes are used to indicate multiples or submultiples of the base unit, except that units for mass are formed by applying the prefix to the symbol g: i.e., Mg not kkg and mg not μkg. Only a single prefix is permitted. Use ns rather than mμs, pF rather than μμF, GW rather than kMW. The first syllable of the prefix retains its stress in compounds; thus, preferred pronunciations are *kil′o-mē-ter* and *mic′ro-mē-ter*, not *ki-lom′e-ter* or *mi-crom′e-ter*.

Pronunciation: **a**go, f**ă**t, f**ā**te, p**ĕ**t, **ē**ven, **go**, p**ĭ**t, p**ô**t, t**ōe**

Factor	Prefix	Symbol	Factor	Prefix	Symbol
10^1	deka ($děk′a$)	da	10^{-1}	deci ($děs′ĭ$) [a]	d
10^2	hecto ($hěk′tō$)	h	10^{-2}	centi ($sěnt′ĭ$) [a]	c
10^3	kilo ($kĭl′ō$)	k	10^{-3}	milli ($mil′ĭ$) [a]	m
10^6	mega ($māg′a$)	M	10^{-6}	micro ($mīk′rō$)	μ
10^9	giga ($gĭg′a$)	G	10^{-9}	nano ($năn′ō$)	n
10^{12}	tera ($těr′a$)	T	10^{-12}	pico ($pēk′ō$)	p
10^{15}	peta ($pět′a$)	P	10^{-15}	femto ($fěm′tō$)	f
10^{18}	exa ($ěx′a$)	E	10^{-18}	atto ($ăt′ō$)	a
10^{21}	zetta ($zět′a$)	Z	10^{-21}	zepto ($zěp′tō$)	z
10^{24}	yotta ($yôt′a$)	Y	10^{-24}	yocto ($yôk′tō$)	y

[a]   Before a vowel these become ($děs′ē$), ($sěnt′ē$), ($mĭl′ē$).

2.1   Realization of the meter

2.1.1   Laser radiations

In principle, the meter should be realized by a determination of the wavelength in vacuum of a plane electromagnetic wave of frequency f, using the relation $\lambda = c/f$ and the defined value of the speed of light in vacuum, $c = 299\,792\,458$ m/s; in practice, it is realized by determining the wavelength of the standing wave of an absorption-stabilized laser from the ratio of the frequency of the laser to the frequency of the cesium clock.

The values listed in the table for the frequencies and wavelengths in vacuum can be used as standards with the uncertainty shown, provided that the specifications of the procedures given by the CIPM are followed. Any of the radiations listed here can be replaced by radiations corresponding to another component of the same transition, or by another radiation, if the difference of the two frequencies is known with sufficient accuracy.

Table 2.3 Realization of the meter using absorption-stabilized laser radiations.

Absorbing Molecule	Transition, component	Frequency f/MHz	Wavelength λ/fm	(unc)[a]
1. CH_4	ν_3,P(7), $(F_2^{(2)})$	88 376 181.608	3 392 231 397.0	1.3
2. $^{127}I_2$	17-1, P(62), o	520 206 808.51	576 294 760.27	0.6
3. $^{127}I_2$	11-5, R(127), i	473 612 214.8	632 991 398.1	1.0
4. $^{127}I_2$	9-2, R(47), o	489 880 355.1	611 970 769.8	1.1
5. $^{127}I_2$	43-0, P(13), a_3 [b]	582 490 603.6	514 673 466.2	1.3

[a] Estimated overall relative uncertainty (3 standard deviations) in parts in 10^9.
[b] Also known as component s.

2.1.2　Radiations of spectral lamps

The wavelength of the radiation corresponding to the transition between levels $2p_{10}$ and $5d_5$ (Paschen notation) of the atom ^{86}Kr is

$$\lambda(^{86}Kr) = 605\,780\,210.2\,\text{fm}, \qquad 1\,\text{m} = 1\,650\,763.731\,\lambda(^{86}Kr).$$

The wavelengths of the atomic transitions given in the table are consistent with this definition with an estimated overall relative uncertainty of 4×10^{-9} when the lamps are operated under specified conditions.[5] These secondary standards provide convenient wavelength markers in the visible region of the spectrum.

Table 2.4 Secondary wavelength standards.

Source	Wavelengths in air $(\lambda/\text{Å})$			
^{86}Kr	6456.2876	6421.0257	5649.5606	4502.3533
^{114}Cd	6438.4685	5085.8203	4799.9104	4678.1487
^{198}Hg	5790.6626	5769.5982	5460.7530	4358.3374

2.2　Derived units

By combining SI base units, it is possible to derive all other units. A derived unit may correspond to more than one physical quantity, but a given physical quantity, although it may be expressed in terms of different units or different equivalent names for the same unit, has a unique dimension and a unique coherent unit within SI.

Several of the most commonly used derived units of SI have been given special names. Preference is given to specific names for certain units in order to facilitate a distinction between quantities with the same dimension. The unit newton meter (N m) may be used for both energy and moment of force (torque), but the unit joule (J) is used only for energy. In the field of ionizing radiation the quantities specific energy, absorbed dose, and dose equivalent each have the dimension $length^2/time^2$; the unit for specific energy in general is joule per kilogram (J/kg), but the unit gray (Gy) is preferred for the quantity dose, and the unit sievert (Sv) should be used exclusively for the quantity dose equivalent.

becquerel, Bq activity of a radionuclide decaying at the rate of 1 transition per second;

coulomb, C quantity of electricity carried in one second by a current of 1 ampere;

farad, F capacitance of a capacitor between the plates of which there appears a potential difference of 1 volt when it is charged by a quantity of electricity of 1 coulomb;

gray, Gy absorbed dose when 1 joule is imparted per kilogram of matter by ionizing radiation. (The gray may also be used to express specific energy imparted, kerma, and absorbed dose index.);

henry, H inductance of a closed circuit in which an electromotive force of 1 volt is produced when the electric current in the circuit varies uniformly at the rate of 1 ampere per second;

hertz, Hz frequency of a periodic phenomenon, the period of which is 1 second;

joule, J work done when the point of application of a force of 1 newton moves a distance of 1 meter in the direction of the force;

lumen, lm luminous flux emitted in a solid angle of 1 steradian by a point source having a uniform intensity of 1 candela;

lux, lx illuminance produced by a luminous flux of 1 lumen uniformly distributed over an area of 1 square meter;

newton, N force that gives to a mass of 1 kilogram an acceleration of 1 meter per second squared;

ohm, Ω electric resistance between two points of a conductor (not being the seat of any electromotive force) when a constant potential difference of 1 volt, applied to these points, produces in the conductor a current of 1 ampere;

pascal, Pa pressure or stress of 1 newton per square meter.

Table 2.5 Derived SI units with special names.

Quantity	Name	Symbol	Expression in terms of base units	Expression in terms of other SI units
Frequency	hertz	Hz	s^{-1}	
Force	newton	N	$m \cdot kg/s^2$	J/m
Pressure	pascal	Pa	$kg/m \cdot s^2$	N/m^2, J/m^3
Energy, work, quantity of heat	joule	J	$m^2 \cdot kg/s^2$	N·m
Power, radiant flux	watt	W	$m^2 \cdot kg/s^3$	J/s
Quantity of electricity, electric charge	coulomb	C	A·s	
Potential difference, electric potential, electromotive force	volt	V	$m^2 \cdot kg/s^3 \cdot A$	W/A, J/C
Capacitance	farad	F	$s^4 \cdot A^2/m^2 \cdot kg$	C/V
Electric resistance	ohm	Ω	$m^2 \cdot kg/s^3 \cdot A^2$	V/A
Electric conductance	siemens	S	$s^3 \cdot A^2/m^2 \cdot kg$	A/V, Ω^{-1}
Magnetic flux	weber	Wb	$m^2 \cdot kg/s^2 \cdot A$	V·s
Magnetic flux density	tesla	T	$kg/s^2 \cdot A$	Wb/m^2
Inductance	henry	H	$m^2 \cdot kg/s^2 \cdot A^2$	Wb/A
Celsius temperature	degree Celsius	°C	K	
Luminous flux	lumen	lm	cd·sr [a]	
Illuminance	lux	lx	$cd \cdot sr/m^2$ [a]	lm/m^2

Units admitted for reasons of safeguarding public health

Quantity	Name	Symbol	Expression in terms of base units	Expression in terms of other SI units
Activity (of a radionuclide)	becquerel	Bq	s^{-1}	
Absorbed dose, specific energy imparted, kerma	gray	Gy	m^2/s^2	J/kg
Dose equivalent [b]	sievert	Sv	m^2/s^2	

[a] The symbol sr must be included here to distinguish luminous flux (lumen) from luminous intensity (candela).

[b] Although the gray and the sievert have the same expression in terms of base units, they measure conceptually distinct quantities. In order to distinguish between the two, the unit J/kg should not be used with dose equivalent.

siemens,   S conductance of a conductor in which a current of 1 ampere is produced by an electric potential difference of one volt;

sievert,   Sv equivalent dose D_{eq} when the product of the dose D of ionizing radiation and the dimensionless factors Q (quality factor) and N (product of any other multiplying factors) stipulated by the International Commission on Radiation Protection is equal to one joule per kilogram. The numerical value of $D_{eq} = Q \cdot N \cdot D$ will differ from that of D depending upon the values of Q and N;

tesla,   T magnetic flux density given by a magnetic flux of 1 weber per square meter;

volt,   V difference of electrical potential between two points of a conducting wire carrying a current of 1 ampere when the power dissipated is 1 watt;

watt,   W power that, in 1 second, gives rise to energy of 1 joule;

weber,   Wb magnetic flux that, linking a circuit of 1 turn, would produce in it an electromotive force of 1 volt if it were reduced to zero at a uniform rate in 1 second.

2.3 Supplementary derived units

Since the value of a quantity with dimension one is invariant to a transformation of units, such a quantity need not be included in the category of base quantities. However, although plane angle is a quantity derived from the ratio of two lengths with the unit "radian" ($1\,\text{rad} = 1\,\text{m/m} = 1$), and solid angle is a quantity derived from the ratio of an area to the square of a length with the unit "steradian" ($1\,\text{sr} = 1\,\text{m}^2/\text{m}^2 = 1$), in some situations these units are treated as base units. In the fields of radiative transport and particle scattering, and notably in photometry and illumination, the unit steradian must be treated as a base unit in order to avoid ambiguity and to distinguish between the candela and the lumen. Similarly, the unit radian must be treated as a base unit in order to distinguish between the unit rad/s and the unit hertz.

2.4 Non-SI systems of units

Although the Système International is the recommended system for representing quantities and units, a great deal of the literature in physics has been, and continues to be, expressed in terms of other systems. It is thus necessary to understand the relationship between SI and these systems if

Table 2.6 Commonly used non-SI units with exactly defined values.

Quantity	Name	Symbol	Definition
Plane angle	degree	°	$1° = \dfrac{\pi}{180}$ rad
	minute (of angle)	′	$1' = \dfrac{1}{60}° = \dfrac{\pi}{10\,800}$ rad
	second (of angle)	″	$1'' = \dfrac{1}{60}' = \dfrac{\pi}{648\,000}$ rad
Time[a]	minute	min	1 min $= 60$ s
	hour	h	1 h $= 60$ min $= 3600$ s
	day	d	1 d $= 24$ h $= 86\,400$ s
Volume	liter	L	1 L $= 1$ dm$^3 = 10^{-3}$ m^3
Mass	metric ton (tonne)	t	1 t $= 1$ Mg $= 1000$ kg

[a] The general symbol for the time unit year (année) is a. The lengths of the various astronomical years are not constant, and there are several different calendar years; when "year" is used as a unit and the numerical precision warrants it, the specific definition should be indicated. The year in astronomical tables is the Julian year (365.25 d).

Table 2.7 Non-SI units with experimentally determined values.

The values given here are approximate; the tables in Chapter 4 give more precise values and uncertainties for these units.

Quantity	Name	Symbol	Definition	Value
Mass	(unified) atomic mass unit	u	$m(^{12}\text{C})/12$	1.6605×10^{-27} kg
Energy	electron volt	eV	$\{e\}$ J [a]	1.6022×10^{-19} J

[a] The symbol $\{e\}$ indicates the numerical value (in coulombs) of the elementary charge e.

the older literature is to be fully utilized. The discussion here is not intended to be a complete review of these systems nor to advance their use; its purpose is to provide a basis for their translation into SI.

2.4.1 Systems with three base quantities

During the 19th century, when it was considered to be a manifestation of mechanical deformations of the ether, electromagnetism was forced into an artificially restrictive three-dimensional framework. As a result, several different systems of electric and magnetic units were developed from

Table 2.8 Units presently sanctioned for use with SI. The use of these units may be discontinued in the future.

Quantity	Unit Name	Symbol	Definition
Length	nautical mile	nm	$1852\,\mathrm{m}$
	ångström	Å	$0.1\,\mathrm{nm} = 10^{-10}\,\mathrm{m}$
	fermi	f, F	$1\,\mathrm{fm} = 10^{-15}\,\mathrm{m}$
Area	barn	b	$100\,\mathrm{fm}^2 = 10^{-28}\,\mathrm{m}^2$
	are	a	$100\,\mathrm{m}^2$
	hectare	ha	$100\,\mathrm{a} = 10^4\,\mathrm{m}^2$
Velocity	knot	kn	$\dfrac{1852}{3600}\,\mathrm{m/s}$
Acceleration	gal[a]	Gal	$1\,\mathrm{cm/s}^2 = 0.01\,\mathrm{m/s}^2$
Pressure	bar	bar	$10^5\,\mathrm{Pa}$
	torr	Torr	$\dfrac{101325}{760}\,\mathrm{Pa}$
Quantity of heat	calorie	cal	
	$1\,\mathrm{cal}_{ST}$		4.1868 J [b]
	$1\,\mathrm{cal}_{th}$		4.184 J [b]
Activity of a radioactive source	curie	Ci	$3.7{\times}10^{10}\,\mathrm{s}^{-1}$
Exposure of x or γ radiations	roentgen	R	$2.58{\times}10^{-4}\,\mathrm{C/kg}$
Absorbed dose	rad	rad, rd [c]	$1\,\mathrm{cGy} = 0.01\,\mathrm{Gy}$
Dose equivalent	rem	rem	$1\,\mathrm{cSv} = 0.01\,\mathrm{Sv}$

[a] The gal is used in geophysics to express the earth's gravitational field; it should not be used as a unit of acceleration other than in this specific sense.

[b] These units are, respectively, the International Steam Table calorie and the thermodynamic calorie.

[c] The symbol rd is used when there is a risk of confusion between rad and the symbol for radian.

the base quantities length, mass, and time. Because such a description of electromagnetism is incomplete, the dimensions of electric and magnetic quantities were dependent on the specific choice of the basic interaction. These systems can be grouped into at least three classes:

a. The "electrostatic" systems in which Coulomb's law for the force between two electric charges,

$$\boldsymbol{F} = k_e \frac{q_1 q_2}{\epsilon r^3} \boldsymbol{r} \tag{1}$$

(with k_e an arbitrarily chosen numerical factor, with the permittivity ϵ defined to be a quantity of dimension one with the value 1 for vacuum),

Table 2.9 Non-rationalized and rationalized systems.

Non-rationalized symmetrical (Gaussian) system with three base quantities (1.c)	Rationalized system with four base quantities

Equations

$$c\nabla \times \boldsymbol{E}^* = -\partial \boldsymbol{B}^*/\partial t$$
$$c\nabla \times \boldsymbol{H}^* = 4\pi \boldsymbol{j}^* + \partial \boldsymbol{D}^*/\partial t$$
$$\nabla \cdot \boldsymbol{D}^* = 4\pi \rho^*$$
$$\nabla \cdot \boldsymbol{B}^* = 0$$
$$\boldsymbol{F} = q^*(\boldsymbol{E}^* + \boldsymbol{v} \times \boldsymbol{B}^*/c)$$
$$w = (\boldsymbol{E}^* \cdot \boldsymbol{D}^* + \boldsymbol{B}^* \cdot \boldsymbol{E}^*)/8\pi$$
$$S = c(\boldsymbol{E}^* \times \boldsymbol{H}^*)/4\pi$$
$$\boldsymbol{E}^* = -(\nabla V^* + (1/c)\partial \boldsymbol{A}^*/\partial t)$$
$$\boldsymbol{B}^* = \nabla \times \boldsymbol{A}^*$$
$$\boldsymbol{D}^* = \epsilon_r \boldsymbol{E}^*$$
$$= \boldsymbol{E}^* + 4\pi \boldsymbol{P}^*$$
$$\boldsymbol{B}^* = \mu_r \boldsymbol{H}^*$$
$$= \boldsymbol{H}^* + 4\pi \boldsymbol{M}^*$$
$$\epsilon_r = 1 + 4\pi \chi_e^*$$
$$\mu_r = 1 + 4\pi \chi_m^*$$

$$\nabla \times \boldsymbol{E} = -\partial \boldsymbol{B}/\partial t$$
$$\nabla \times \boldsymbol{H} = \boldsymbol{j} + \partial \boldsymbol{D}/\partial t$$
$$\nabla \cdot \boldsymbol{D} = \rho$$
$$\nabla \cdot \boldsymbol{B} = 0$$
$$\boldsymbol{F} = q(\boldsymbol{E} + \boldsymbol{v} \times \boldsymbol{B})$$
$$w = \tfrac{1}{2}(\boldsymbol{E} \cdot \boldsymbol{D} + \boldsymbol{B} \cdot \boldsymbol{E})$$
$$S = \boldsymbol{E} \times \boldsymbol{H}$$
$$\boldsymbol{E} = -(\nabla V + \partial \boldsymbol{A}/\partial t)$$
$$\boldsymbol{B} = \nabla \times \boldsymbol{A}$$
$$\boldsymbol{D} = \epsilon \boldsymbol{E} = \epsilon_0 \epsilon_r \boldsymbol{E}$$
$$= \epsilon_0 \boldsymbol{E} + \boldsymbol{P}$$
$$\boldsymbol{B} = \mu \boldsymbol{H} = \mu_0 \mu_r \boldsymbol{H}$$
$$= \mu_0 (\boldsymbol{H} + \boldsymbol{M})$$
$$\epsilon_r = 1 + \chi_e$$
$$\mu_r = 1 + \chi_m$$

Physical constants

$$\alpha = e^{*2}/\hbar c$$
$$a_0 = \hbar^2/m_e e^{*2}$$
$$hcR_\infty = e^{*2}/2a_0$$
$$r_e = e^{*2}/m_e c^2$$
$$\mu_B^* = e^* \hbar/2m_e c$$
$$\omega_L = (q^*/2mc)B^*$$
$$\gamma^* = gI(e^*/mc)$$

$$\alpha = e^2/4\pi \epsilon_0 \hbar c = \mu_0 c e^2/2h$$
$$a_0 = 4\pi \epsilon_0 \hbar^2/m_e e^2$$
$$hcR_\infty = e^2/8\pi \epsilon_0 a_0$$
$$r_e = \mu_0 e^2/m_e$$
$$\mu_B = e\hbar/2m_e$$
$$\omega_L = (q/2m)B$$
$$\gamma = gI(e/m)$$

was used as the basis for defining the unit of charge. For a physical medium ϵ is equal to the relative permittivity of SI. In this system charge has the dimension $L^{3/2}M^{1/2}T^{-1}$ and electric current has the dimension $L^{3/2}M^{1/2}T^{-2}$.

b. The "electromagnetic" systems in which Ampère's law for the force between two electric current elements,

$$d^2 \boldsymbol{F} = k_m \mu \frac{i_1 d\boldsymbol{l}_1 \times (i_2 d\boldsymbol{l}_2 \times \boldsymbol{r})}{r^3}, \tag{2}$$

(with k_m similarly arbitrary, and with the permeability μ defined to be a quantity of dimension one with the value 1 for a vacuum) was used as the basis for defining the unit of current. In this system, current has the dimension $L^{1/2}M^{1/2}T^{-1}$ and electric charge has the dimension $L^{1/2}M^{1/2}$.

c. The "symmetrical" Gaussian system using electric quantities (including electric current) from system (a) and magnetic quantities from system (b).

The ratio of the dimension of charge or current in the system (a) to its dimension in system (b) is LT^{-1}, which is reflected in the appearance of explicit factors of the square of the speed of light in many of the relations involving electromagnetic forces (proportional to the second power of charge). In system (c) the first power of the speed of light appears in equations relating electric and magnetic quantities.

The choices $k_e = 1$ in eq. (1) and $k_m = 1$ in eq. (2) lead to the appearance of the factors 2π and 4π in equations in situations that involve plane geometry, and to their absence in situations that have cylindrical or spherical symmetry where these factors might normally be expected. These systems are therefore called "irrational" or "non-rationalized," since, if the factors k_e and k_m are set equal to $1/4\pi$ in eqs. (1) and (2), respectively (recognizing the spherical symmetry of these equations), the factors of 2π and 4π then appear explicitly only in those equations where they would be expected from the geometry of the system. In this form the equations are said to be "rationalized."

2.4.2   Systems with four base quantities

In SI, as in its older relative, the MKSA system, the three base quantities of mechanics are augmented by a fourth base quantity, which is taken to be electric current, and in the Système International eqs. (1) and (2) are rationalized ($k_e = k_m = 1/4\pi$). As a result, the ampere is an independent base unit, not simply a convenient name for the coherent unit for a derived quantity involving fractional powers of the mechanical base units. As a result, permeability μ and permittivity ϵ are dimensional physical quantities, and if electrostatics and electrodynamics are to be coherent, avoiding the explicit introduction of the factor c asymmetrically into the expressions for electric and magnetic quantities, ϵ_o and μ_o must satisfy the condition

$$\epsilon_o \mu_o c^2 = 1.$$

In SI the permeability of vacuum μ_o is defined to have the value

$$\mu_o = 4\pi \times 10^{-7} \, \text{N/A}^2 = 4\pi \times 10^{-7} \, \text{H/m} = 4\pi \times 10^{-7} \, \text{T·m/A}.$$

Relations between quantities in different systems: The basic equations between quantities in the non-rationalized symmetrical (Gaussian) system (c) and the corresponding equations in the rationalized four-dimensional system are given in the table. In order to distinguish the physical quantities in the two systems, those in the three-dimensional system are indicated with an asterisk (*) when they differ from the corresponding quantities of the rationalized four-dimensional system. The relationships between the two sets of quantities are determined by setting $X^* = a_X X$ in the first column and comparing the resultant equations with the corresponding ones in the second column. These substitutions lead to:

$$\frac{E^*}{E} = \frac{V^*}{V} = \frac{\rho}{\rho^*} = \frac{Q}{Q^*} = \frac{j}{j^*} = \frac{I}{I^*} = \frac{P}{P^*} = \sqrt{4\pi\epsilon_\circ} \,,$$

$$\frac{D^*}{D} = \sqrt{4\pi/\epsilon_\circ} \,,$$

$$\frac{H^*}{H} = \sqrt{4\pi\mu_\circ} \,,$$

$$\frac{B^*}{B} = \frac{A^*}{A} = \frac{M}{M^*} = \sqrt{4\pi/\mu_\circ} \,,$$

$$\frac{\chi_e}{\chi_e^*} = \frac{\chi_m}{\chi_m^*} = 4\pi \,.$$

The cgs systems of electrical units: The cgs "electrostatic" system of units (esu) forms a coherent system by combining the mechanical units centimeter, gram, and second with the "electrostatic" system of quantities of (a). In its less common form as a four-dimensional system, the electrostatic unit of charge (sometimes called the franklin; symbol, Fr) is introduced, and the permittivity of vacuum is set equal to

$$\epsilon_\circ = 1 \, \text{Fr}^2 \, \text{dyn}^{-1} \, \text{cm}^{-2} \,.$$

Other units may then be derived using the usual rules for constructing a coherent set of units from a set of base units.

The cgs "electromagnetic" system of units (emu) forms a coherent system with the three-dimensional electromagnetic system of quantities of (b). In its four-dimensional form, the fourth base unit is taken to be the current unit abampere (symbol, abamp), or biot (symbol, Bi), by defining the permeability of vacuum to be

$$\mu_\circ = 1 \, \text{g cm s}^{-2} \, \text{abamp}^{-2} \,.$$

The force between two parallel, infinitely long wires, 1 cm apart in vacuum, each carrying a current of 1 abamp, is 2 dyn per cm of length.

The "mixed," "symmetrized," or "Gaussian" cgs units, consisting of the set of electric units of the esu system and the magnetic units of the emu system, form a coherent system of units when used with the three-dimensional "symmetrical system" or "Gaussian system" of equations (c).

Table 2.10 cgs base units and derived units with special names.

Quantity	Name	Symbol	Expression in terms of base units	Value
Length	centimeter	cm		10^{-2} m
Mass	gram	g		10^{-3} kg
Time	second	s		
Force	dyne	dyn	cm g s^{-2}	10^{-5} N
Energy	erg	erg	$\text{cm}^2\,\text{g}^{-2}$	10^{-7} J
(Dynamic) viscosity	poise	P	$\text{cm}^{-1}\,\text{g s}^{-1}$	0.1 Pa·s
Kinematic viscosity	stokes	St	$\text{cm}^2\,\text{s}^{-1}$	10^{-4} m^2/s
Acceleration of free fall[a]	gal	Gal	cm s^{-2}	0.01 m/s^2

[a] The gal is a unit of acceleration only for the strength of the earth's gravitational field.

Table 2.11 cgs units in photometry.   Units derived from cm, g, s, cd and sr have been given special names and symbols.

Quantity	Name	Symbol	Expression	Value
Luminance	stilb	sb	$\text{cm}^{-2}\,\text{cd}$	10^4 cd/m^2
Illuminance	phot	ph	$\text{cm}^{-2}\,\text{cd sr}$	10^4 lux

Table 2.12 cgs magnetic units with special names.

Quantity	Name	Symbol	Dimension[a]	Equivalence between cgs units and corresponding SI units
H^*	oersted	Oe	$L^{-1/2}\,M^{1/2}\,T^{-1}$	$\dfrac{1}{4\pi}$ abamp/cm $= \dfrac{10^{-4}}{\mu_0}$ T
B^*	gauss	G, (Gs)	$L^{-1/2}\,M^{1/2}\,T^{-1}$	10^{-4} T
Φ^*	maxwell	Mx	$L^{3/2}\,M^{1/2}\,T^{-1}$	10^{-8} Wb
F_m^*	gilbert	Gi, (Gb)	$L^{1/2}\,M^{1/2}\,T^{-1}$	$\dfrac{1}{4\pi}$ abamp $= \dfrac{10^{-6}}{\mu_0}$ T·m

[a] L = length; M = mass; T = time.

Table 2.13 Conversion factors for cgs electrical units. An exact equality between the four-dimensional SI and the three-dimensional cgs systems is not strictly possible. In cgs electromagnetic systems μ_o is equal either to 1 or to 4π; in electrostatic systems ϵ_o is equal either to 1 or to $1/4\pi$. μ_o and ϵ_o have dimension one; current and charge are then derived quantities.

$$\xi \equiv 2.997\,924\,58$$
$$\xi^2 = 8.987\,551\,79 \qquad 1/\xi = 0.333\,564\,095 \qquad 1/\xi^2 = 0.111\,265\,006$$

Quantity	emu		esu	
Capacitance	abfarad	$= 10^9$ F	statfarad	$= (10^{-11}/\xi^2)$ F
Charge	abcoulomb	$= 10$ C	statcoulomb[a]	$= (10^{-9}/\xi)$ C
Conductance	abmho	$= 10^9$ S	statmho	$= (10^{-11}/\xi^2)$ S
Current	abampere[b]	$= 10$ A	statampere	$= (10^{-9}/\xi)$ A
Electrical potential	abvolt	$= 10^{-8}$ V	statvolt	$= 100\xi$ V
Inductance	abhenry	$= 10^{-9}$ H	stathenry	$= 10^{11}\xi^2$ H
Resistance	abohm	$= 10^{-9}$ Ω	statohm	$= 10^{11}\xi^2$ Ω

[a]   The statcoulomb is also known as the franklin (Fr).
[b]   The abampere is also known as the biot (Bi).

Special names and symbols have been given to four of the magnetic emu or Gaussian cgs units. In evaluating the relationship between a cgs unit and an SI rationalized unit one must include not only the transformation of the *quantities* given in the preceding section but also the transformation of the *units* from centimeter and gram to meter and kilogram. In addition, the relationship between a four-dimensional electrical unit and its corresponding three-dimensional unit includes the quantity μ_o, with the recognition that its value is unity in the unrationalized three-dimensional system.

Table 2.14 Conversion factors to SI units. The conversion factor for a compound unit is generally not given here if it may easily be derived from simpler conversions; e.g., the conversion factors for "ft/s" to "m/s" or "ft/s^2" to "m/s^2" are not given, since they may be obtained from the conversion factor for "ft." Values are given, generally, to five or six significant digits except for exact values, indicated by $\equiv$. Physical constants used as units will be found in Chapter 4.

1. Angle

1 second ($''$)	$= 4.48481 \times 10^{-6}$ rad
1 minute ($'$)	$= 2.9089 \times 10^{-4}$ rad
1 degree ($^\circ$)	$= 0.0174\,532$ rad
1 rad	$= 206\,264.8''$

Table 2.14 *Continued.*

2. *Area*

1 barn (b)	$\equiv 10^{-28}\,\text{m}^2$
1 circular mil	$= 5.0671 \times 10^{-10}\,\text{m}^2$
1 in.^{2}	$\equiv 6.4516 \times 10^{-4}\,\text{m}^2$
1 ft^2	$\equiv 0.092\,903\,04\,\text{m}^2$
1 yd^2	$\equiv 0.836\,127\,36\,\text{m}^2$
1 are	$\equiv 100\,\text{m}^2$
1 acre [43560 (statute ft)^2]	$= 4046.873\,\text{m}^2$
1 hectare	$\equiv 10\,000\,\text{m}^2$
1 mi^2	$= 2.5900 \times 10^6\,\text{m}^2$

3. *Concentration, Density*

1 grain/gal (US)	$= 0.017\,118\,\text{kg/m}^3$
1 lb/ft^3	$= 16.018\,\text{kg/m}^3$
1 lb/gal (US)	$= 119.83\,\text{kg/m}^3$
1 short ton/yd^3	$= 1186.6\,\text{kg/m}^3$
1 long ton/yd^3	$= 1328.9\,\text{kg/m}^3$
1 oz(avdp)/in^3	$= 1730.0\,\text{kg/m}^3$
1 lb/in^3	$= 27\,680\,\text{kg/m}^3$

4. *Energy*

1 erg	$\equiv 10^{-7}\,\text{J}$
1 ton TNT (equivalent)	$\equiv 4.184 \times 10^9\,\text{J}$
1 ft·lbf	$= 1.3558\,\text{J}$
1 cal$_{th}$ (thermochemical calorie)	$\equiv 4.184\,\text{J}$
1 cal$_{15}$ (15 °C calorie)	$\equiv 4.1855\,\text{J}$
1 cal$_{ST}$ (International Steam Table calorie)	$\equiv 4.1868\,\text{J}$
1 watt second (W·s)	$\equiv 1\,\text{J}$
1 watt hour (W·h)	$\equiv 3600\,\text{J}$
1 therm (EC)	$\equiv 1.055\,06 \times 10^8\,\text{J}$
1 therm (U.S.)	$\equiv 1.054\,804 \times 10^8\,\text{J}$
1 Btu$_{th}$	$= 1054.350\,\text{J}$
1 Btu$_{15}$	$= 1054.728\,\text{J}$
1 Btu$_{ST}$	$\equiv 1055.055\,852\,62\,\text{J}$
1 quad $\equiv 10^{15}$ Btu	$\approx 10^{18}\,\text{J} = 1\,\text{EJ}$

5. *Force*

1 dyne	$\equiv 10^{-5}\,\text{N}$
1 ounce-force	$= 0.27801\,\text{N}$
1 pound-force	$= 4.4482\,\text{N}$
1 kilogram-force	$\equiv 9.80665\,\text{N}$
1 kip (1000 lbf)	$= 4448.2\,\text{N}$
1 ton-force (2000 lbbf)	$= 8896.4\,\text{N}$

6. *Heat Transfer*

1 Btu/lb °F $\equiv$ 1 cal$_{ST}$/g °C		$\equiv 4186.8\,\text{J/kg K}$
Thermal conductivity:	1 Btu ft/h ft^2 °F	$= 1.730\,735\,\text{W m}^{-1}\,\text{K}^{-1}$
	1 Btu in/s ft^2 °F	$= 519.2204\,\text{W m}^{-1}\,\text{K}^{-1}$
Thermal resistance:	1 °F h ft^2/Btu	$= 0.176\,11\,\text{K m}^2/\text{W}$
Thermal resistivity:	1 °F h ft^2/Btu·in	$= 6.933\,47\,\text{K m/W}$

Table 2.14 *Continued.*

7. Length

1 fermi	$\equiv 10^{-15}$ m $= 1$ fm
1 ångström (Å)	$\equiv 10^{-10}$ m
1 microinch	$\equiv 2.54 \times 10^{-8}$ m
1 mil	$\equiv 2.54 \times 10^{-5}$ m
1 point (pt) [0.013 837 in.] [a]	$= 0.351\,46$ mm
1 pica (12 pt)	$= 4.2175$ mm
1 inch (in.)	$\equiv 0.0254$ m
1 hand (4 in.)	$\equiv 0.1016$ m
1 foot (12 in.)	$\equiv 0.3048$ m
1 statute foot [(1200/3937) m] [b]	$= 0.304\,8006$ m
1 yard (yd)	$\equiv 0.9144$ m
1 fathom (6 statute ft)	$= 1.8288$ m
1 rod (16.5 statute ft)	$= 5.0292$ m
1 chain (4 rod)	$= 20.1168$ m
1 furlong (10 chain)	$= 201.168$ m
1 mile (8 furlong, 5280 ft)	$\equiv 1609.344$ m
1 statute mile (8 furlong, 5280 statute ft)	$= 1609.3472$ m
1 nautical mile [b]	$\equiv 1852$ m

8. Light

1 foot-candle	$= 10.764$ lx
1 phot	$\equiv 10\,000$ lx
1 cd/in^2	$= 1550.003$ cd/m^2
1 foot-lambert	$= 3.426\,26$ cd/m^2
1 lambert	$= 3183.10$ cd/m^2
1 stilb	$\equiv 10\,000$ cd/m^2

9. Mass

1 pound (avdp) (lb) (7000 gr)	$\equiv 0.453\,592\,37$ kg
1 pound (troy) (5760 gr)	$\equiv 0.373\,241\,7216$ kg
1 grain (gr)	$\equiv 64.79891$ mg
1 scruple (20 gr)	$= 1.29\,60$ g
1 pennyweight (24 gr)	$= 1.5552$ g
1 dram (60 gr)	$= 3.8879$ g
1 ounce (avdp) (437.5 gr)	$= 28.3495$ g
1 ounce (troy) (480 gr)	$= 31.1035$ g
1 carat (metric)	$= 0.2$ g
1 stone (14 lb)	$= 6.350\,29$ kg
1 slug	$= 14.5939$ kg
1 hundredweight (long)	$= 50.8023$ kg
1 ton (short)	$= 907.185$ kg
1 ton (long)	$= 1016.047$ kg

10. Power   (*1* Btu$_{ST}$=*1.000 669* Btu$_{th}$)

1 cal$_{ST}$/s	$\equiv 4.1868$ W
1 cal$_{th}$/s	$\equiv 4.184$ W
1 erg/s	$\equiv 10^{-7}$ W
1 ft·lbf/h	$= 3.7662 \times 10^{-4}$ W

Table 2.14 *Continued.*

1 Btu_{th}/h	$= 0.292\,875$ W
1 Btu_{ST}/h	$= 0.293\,071$ W
1 metric horsepower (force de cheval)	$= 735.50$ W
1 horsepower (550 ft·lbf/s)	$= 745.70$ W
1 electric horsepower	$\equiv 746$ W

11. Pressure, Stress   (standard atmosphere $\equiv 101\,325$ Pa)

1 $dyne/cm^2$	$\equiv 0.1$ Pa
1 torr [(101325/760) Pa]	$= 133.3224$ Pa
1 N/cm^2	$\equiv 10\,000$ Pa
1 bar	$\equiv 100\,000$ Pa
1 lbf/ft^2	$= 47.880$ Pa
1 cm water (4 °C)	$= 98.0637$ Pa
1 gmf/cm^2	$\equiv 98.0665$ Pa
1 cm of mercury (0 °C)	$= 1333.224$ Pa
1 in of water (39.2 °F)	$= 249.08$ Pa
1 in of mercury (39.2 °F)	$= 3386.4$ Pa
1 lbf/in^2 (psi)	$= 6894.8$ Pa
1 kgf/cm^2	$\equiv 98\,066.5$ Pa

12. Torque

1 dyne·cm	$\equiv 10^{-7}$ N m
1 kgf·m	$\equiv 9.80665$ N m
1 ozf·in	$= 0.007\,0616$ N m
1 lbf·in	$= 0.112\,985$ N m
1 lbf·ft	$= 1.35582$ N m

13. Viscosity

1 poise	$\equiv 0.1$ Pa s
1 lb/ft s	$= 1.4882$ Pa s
1 lb/ft h	$= 4.1338 \times 10^{-4}$ Pa s
1 rhe	$\equiv 10$ $Pa^{-1}s^{-1}$
1 stokes	$\equiv 10^{-4}$ m^2/s
1 ft^2/s	$= 0.092\,903$ m^2/s
1 slug/ft s	$= 47.880$ Pa s
1 $lbf·s/ft^2$	$= 47.880$ Pa s
1 $lbf·s/in^2$	$= 6894.8$ Pa s

14. Volume

1 stere	$\equiv 1$ m^3
1 liter	$\equiv 0.001$ m^3
1 ft^3	$= 0.0283\,168$ m^3
1 in^3	$= 1.6387 \times 10^{-5}$ m^3
1 board foot	$= 2.3597 \times 10^{-3}$ m^3
1 acre·foot	$= 1233.48$ m^3
1 dram (U.S. fluid)	$= 3.6967 \times 10^{-6}$ m^3
1 teaspoon (tsp)	$= 4.9288 \times 10^{-6}$ m^3
1 tablespoon (tbsp)	$= 1.4787 \times 10^{-5}$ m^3
1 ounce (U.S. fluid)	$= 2.9574 \times 10^{-5}$ m^3
1 gill (U.S.)	$= 1.1829 \times 10^{-4}$ m^3

Table 2.14 *Continued.*

1 pint (U.S. fluid)		$= 4.7318 \times 10^{-4}\, \mathrm{m}^3$
1 quart (U.S. fluid)		$= 9.4635 \times 10^{-4}\, \mathrm{m}^3$
1 gallon (U.S. liquid)	[231 in^3]	$= 3.7854 \times 10^{-3}\, \mathrm{m}^3$
1 wine barrel (bbl)	[31.5 gal (US)]	$= 0.119\,240\, \mathrm{m}^3$
1 ounce (U.K. fluid)		$= 2.8413 \times 10^{-5}\, \mathrm{m}^3$
1 gill (U.K.)		$= 1.4206 \times 10^{-4}\, \mathrm{m}^3$
1 gallon (U.K.)		$\equiv 4.546\,09 \times 10^{-3}\, \mathrm{m}^3$
		$= 1.200\,950\, \mathrm{gal\,(U.S.)}$
1 pint (U.S. dry)		$= 5.5061 \times 10^{-4}\, \mathrm{m}^3$
1 quart (U.S. dry)		$= 1.1012 \times 10^{-3}\, \mathrm{m}^3$
1 gallon (U.S. dry)		$= 4.4049 \times 10^{-3}\, \mathrm{m}^3$
1 peck		$= 8.8098 \times 10^{-3}\, \mathrm{m}^3$
1 bushel (U.S.)	[2150.42 in^3]	$= 3.5239 \times 10^{-2}\, \mathrm{m}^3$

[a]   Typographers' definition, (*1886*).
[b]   1 ft $\equiv 0.999\,998$ statute ft.

Table 2.15 Temperature scale conversions.   Both the Kelvin scale and the Rankine scale measure the same physical quantity, *thermodynamic temperature*, but in different units. Strictly speaking, a temperature on the Celsius or Fahrenheit *scale* (particularly when referred to a scale such as ITS-90) is a number rather than a physical quantity.

Triple point of natural water:	$T_{\mathrm{tp}} = 273.16\,\mathrm{K}$
Absolute to Rankine:	$T_{\mathrm{R}} = \dfrac{9}{5}(T/\mathrm{K})$
Absolute to Celsius:	$t = (T/\mathrm{K}) - 273.15$
Absolute to Fahrenheit:	$t_{\mathrm{F}} = \dfrac{9}{5}(T/\mathrm{K}) - 459.67$
Celsius to Fahrenheit:	$t_{\mathrm{F}} = \dfrac{9}{5}t + 32$
	$5(t_{\mathrm{F}} + 40) = 9(t + 40)$

References

[1]   *The International System of Units (SI)*, Natl. Inst. Stand. Technol., Spec. Publ. 330 (1991 edition), (U.S. Government Printing Office, Washington, D.C., 1991).

[2]   B. N. Taylor, *Guide for the Use of the International System of Units, 1995 edition*, Natl. Inst. Stand. Technol., Spec. Publ. 811, (U.S. Government Printing Office, Washington, D.C., 1995).

[3]   B. N. Taylor, *Interpretation of the SI for the United States and Metric Conversion Policy for Federal Agencies*, Natl. Inst. Stand. Technol., Spec. Publ. 814, (U.S. Government Printing Office, Washington, D.C., 1991).

[4]   ANSI/IEEE, *American National Standard for Metric Practice*, Std 268-1991, IEEE Standards Coordinating Committee 14 (IEEE, New York, 1992).

[5]   CIPM, *Procès-Verbaux CIPM*, 49th session, 1960, pp 71–72; *Comptes Rendues, 11th GCPM*, 1960, p. 85.

3

RECOMMENDED SYMBOLS FOR PHYSICAL QUANTITIES

3.1 Letter symbols

This chapter presents a listing of the most commonly used symbols for physical quantities. The symbols for *physical constants* (and their recommended values) are given in the following chapter. The list of symbols given here is not intended to be exhaustive, and the absence of a symbol should not of itself prohibit its use. Many of the symbols listed are general; they may be made more specific by adding superscripts, subscripts, or other decorations and by using both lower and upper case forms. An expression given with the name of a symbol should be considered to be descriptive rather than definitive.

The dimension of a quantity is given in terms of the base quantities of SI:

length	L	temperature	Θ
mass	M	amount of substance	N
time	T	luminous intensity	J
electric current	I		

(Strictly, generalized coordinates are numerical markers with dimension 1; the dimensions usually associated with coordinates are contained in the metric tensor of the geometry. Generalized momenta then have the dimension of energy. This is indicated by an asterisk.)

		Dimension	SI Unit
a	annihilation operator	1	
	relative chemical activity	1	
	specific activity	$M^{-1}T^{-1}$	Bq/kg
	thermal diffusivity: $\lambda/\rho c_p$	L^2T^{-1}	m^2/s

		Dimension	SI Unit
$a^\dagger$	creation operator	1	
a	acceleration	LT^{-2}	m/s^2
b	breadth, impact parameter	L	m
	mobility ratio: μ_n/μ_p	1	
	phonon annihilation operator	1	
b_B	molality of solute B	$M^{-1}N$	mol/kg
$b^\dagger$	phonon creation operator	1	
b	Burgers' vector	L	m
c	concentration: $c = n/V$	$L^{-3}N$	mol/m^3
	specific heat capacity	$L^2T^{-2}\Theta^{-1}$	$J/(kg{\cdot}K)$
	speed	LT^{-1}	m/s
	speed of light, speed of sound	LT^{-1}	m/s
c_{ijkl}	elasticity tensor: $\tau_{ij} = c_{ijkl}\epsilon_{lk}$	$L^{-1}MT^{-2}$	Pa, N/m^2
$\bar{c}$	average speed	LT^{-1}	m/s
$\hat{c}$	most probable speed	LT^{-1}	m/s
$\langle c \rangle$	average speed	LT^{-1}	m/s
c	velocity, average velocity	LT^{-1}	m/s
$\langle c \rangle$	average velocity	LT^{-1}	m/s
d	relative density	1	
d	diameter, distance, thickness	L	m
	lattice plane spacing	L	m
e	linear strain	1	
	specific energy	L^2T^{-2}	J/kg
e	polarization vector	1	
f	focal distance	L	m
	frequency	T^{-1}	Hz
f_B	activity coefficient of B (in a mixture)	1	
g	acceleration of free fall	LT^{-2}	m/s^2
	g-factor: $\mu/I\mu_N$	1	
	statistical weight (degeneracy)	1	
h	height	L	m
	heat transfer coefficient	$MT^{-3}\Theta^{-1}$	$W/(m^2{\cdot}K)$
h_1, h_2, h_3			
h, k, l	Miller indices	1	
j	electric current density	$L^{-2}I$	A/m^2
j_i	total angular momentum quantum number	1	
k	angular wave number	L^{-1}	m^{-1}
k_T	thermal diffusion ratio	1	

		Dimension	SI Unit
k	angular wave vector, propagation vector	L^{-1}	m^{-1}
l	length, mean free path	L	m
l_i	orbital angular momentum quantum number	1	
l_e	mean free path of electrons	L	m
l_{ph}	mean free path of phonons	L	m
m	mass	M	kg
	molality of solution	$M^{-1}N$	mol/kg
m^*	effective mass	M	kg
m_a	atomic mass	M	kg
m_i	magnetic quantum number	1	
m_r	reduced mass: $m_1 m_2/(m_1 + m_2)$	M	kg
m_u	atomic mass unit: $\frac{1}{12} m_a(^{12}C)$	M	kg
m_N	nuclear mass	M	kg
m	magnetic dipole moment	$L^2 I$	$A \cdot m^2$
n	amount of substance	N	mol
	electron density (conduction band)	L^{-3}	m^{-3}
	number density of particles	L^{-3}	m^{-3}
	order of reflection	1	
	principal quantum number	1	
	refractive index	1	
n_a	acceptor number density	L^{-3}	m^{-3}
n_d	donor number density	L^{-3}	m^{-3}
n_i	intrinsic number density: $(np)^{1/2}$	L^{-3}	m^{-3}
n_n	electron number density	L^{-3}	m^{-3}
n_p	hole number density	L^{-3}	m^{-3}
p	acoustic pressure, pressure	$L^{-1}MT^{-2}$	$Pa, J/m^3$
	hole density (conduction band)	L^{-3}	m^{-3}
p	electric dipole moment	LTI	$C \cdot m$
	momentum: mv	LMT^{-1}	$kg \cdot m/s$
p, p_i	generalized momentum: $\partial L/\partial q_i$	varies	*
q	electric charge	TI	$A \cdot s$
	flow rate	$L^3 T^{-1}$	m^3/s
q_m	mass flow rate	MT^{-1}	kg/s
q_D	Debye angular wave number	L^{-1}	m^{-1}
q	(phonon) propagation vector	L^{-1}	m^{-1}
q, q_i	generalized coordinate	varies	*

		Dimension	SI Unit
r	distance, radius	L	m
	molar ratio of solute	1	
$\boldsymbol{r}$	position vector	L	m
s	path length	L	m
	long range order parameter	1	
	symmetry number	1	
s_i	spin quantum number	1	
s_{klji}	compliance tensor: $\epsilon_{kl} = s_{klji}\tau_{ij}$	$M^{-1}LT^2$	m^2/N
$\boldsymbol{s}$	position vector	L	m
t	time	T	s
	temperature	Θ	K
u	average speed	LT^{-1}	m/s
	electromagnetic energy density	$L^{-1}MT^{-2}$	J/m^3
$\boldsymbol{u}$	displacement vector	L	m
	velocity	LT^{-1}	m/s
v	specific volume	$L^{-3}M$	m^3/kg
	speed: ds/dt	LT^{-1}	m/s
	vibrational quantum number	1	
	volume	L^3	m^3
v_{dr}	drift velocity (speed)	LT^{-1}	m/s
$\bar{v}$	average speed	LT^{-1}	m/s
$\hat{v}$	most probable speed	LT^{-1}	m/s
$\langle v \rangle$	average speed	LT^{-1}	m/s
$\boldsymbol{v}$	velocity	LT^{-1}	m/s
$\boldsymbol{v}_\circ$	average velocity	LT^{-1}	m/s
$\langle \boldsymbol{v} \rangle$	average velocity	LT^{-1}	m/s
w	electromagnetic energy density	$L^{-1}MT^{-2}$	J/m^3
	mass fraction	1	
x	molar fraction	1	
z	ionic charge number	1	
	reduced activity: $[2\pi mkT/h^2]^{3/2}\lambda$	1	
A	activity (radioactivity)	T^{-1}	s^{-1}, Bq
	area	L^2	m^2
	chemical affinity	$L^2MT^{-2}N^{-1}$	J/mol
	Helmholtz free energy	L^2MT^{-2}	J
	nucleon number, mass number	1	
	Richardson constant: $j = AT^2 \exp(-\Phi/kT)$	$M^{-2}I\Theta^{-2}$	$A/(m^2 \cdot K^2)$
A_H	Hall coefficient	$M^3T^{-1}I^{-1}$	m^3/C
A_r	relative atomic mass: m_a/m_u	1	

		Dimension	SI Unit
A	magnetic vector potential	$LMT^{-2}I^{-1}$	Wb/m, N/A
B	susceptance	$L^{-2}M^{-1}T^3I^2$	S
B	magnetic flux density	$MT^{-2}I^{-1}$	T, N/(A·m)
C	capacitance	$L^{-2}M^{-1}T^4I^2$	A·s/V
	heat capacity	$L^2MT^{-2}\Theta^{-1}$	J/K
D	Debye–Waller factor	1	
	diffusion coefficient	L^2T^{-1}	m²/s
D_{td}	thermal diffusion coefficient	L^2T^{-1}	m²/s
D	electric displacement	$L^{-2}TI$	C/m²
E	electromotive force	$L^2MT^{-3}I^{-1}$	V, J/C
	energy	L^2MT^{-2}	J
	irradiance	MT^{-3}	J/m²
	illuminance	$L^{-2}J$	lm/m², lx
	Young's modulus	$L^{-1}MT^{-2}$	Pa, N/m²
E_a	acceptor ionization energy	L^2MT^{-2}	J, eV
E_{ab}	thermoelectromotive force	$L^2MT^{-3}I^{-1}$	V
E_d	donor ionization energy	L^2MT^{-2}	J, eV
E_g	energy gap	L^2MT^{-2}	J, eV
E_k	kinetic energy	L^2MT^{-2}	J
E_p	potential energy	L^2MT^{-2}	J
E_F	Fermi energy	L^2MT^{-2}	J, eV
E	electric field	$LMT^{-3}I^{-1}$	N/C, V/m
$\mathcal{E}$	electromotive force	$L^2MT^{-3}I^{-1}$	V, J/C
F	hyperfine quantum number	1	
	Helmholtz free energy	L^2MT^{-2}	J
F_m	magnetomotive force	I	A
F	force	LMT^{-2}	N, kg·m/s²
G	conductance	$L^{-2}M^{-1}T^3I^2$	S
	Gibbs free energy	L^2MT^{-2}	J
	shear modulus	$L^{-1}MT^{-2}$	Pa, N/m²
G	reciprocal lattice vector, $G \cdot R = 2\pi m$	L^{-1}	m⁻¹
	enthalpy	L^2MT^{-2}	J
H_c	superconductor critical field strength	$L^{-1}I$	A/m
H	angular impulse: $\int M\,dt$	L^2MT	N·m·s
	magnetic field strength	$L^{-1}I$	A/m
I	electric current	I	A
	luminous intensity	J	cd
	moment of inertia	L^2M	kg·m²

		Dimension	SI Unit
	nuclear spin quantum number (atomic physics)	1	
	radiant intensity	L^2MT^{-2}	W/sr
I	impulse: $\int F\,dt$	LMT^{-1}	N·s
J	action integral: $\oint p\,dq$	L^2MT^{-1}	J·s
	electric current density	$L^{-2}I$	A/m²
	exchange integral	L^2MT^{-2}	J
	nuclear spin quantum number (nuclear physics)	1	
	rotational quantum number	1	
	total angular momentum quantum number	1	
K	bulk modulus	$L^{-1}MT^{-2}$	Pa, N/m²
	equilibrium constant	1	
	heat transfer coefficient	$MT^{-3}\Theta^{-1}$	W/(m²·K)
	kerma (kinetic energy released in matter)	L^2T^{-2}	Gy, J/kg
	kinetic energy	L^2MT^{-2}	J
	luminous efficacy	$L^{-2}M^{-1}T^3J$	lm/W
	relative permittivity	1	
	rotational quantum number	1	
L	length	L	m
	Lorenz coefficient: $\lambda/\sigma T$	$L^4M^2T^{-6}I^{-2}\Theta^{-2}$	V²/K²
	luminance	$M^{-2}J$	cd/m²
	orbital angular momentum quantum number	1	
	self-inductance	$L^2MT^{-2}I^{-2}$	H, Wb/A
L_p	sound pressure level	1	bel, db
L_N	loudness level	1	bel, dB
L_W	sound power level	1	bel, dB
L_{12}	mutual inductance	$L^2MT^{-2}I^{-2}$	H, Wb/A
L	angular momentum: $r \times p$	L^2MT^{-1}	kg·m²/s, J·s
M	magnetic quantum number	1	
	molar mass	MN^{-1}	kg/mol
	mutual inductance	$L^2MT^{-2}I^{-2}$	H, Wb/A
	radiant existance	MT^{-3}	W/m²
	torque, bending moment	L^2MT^{-2}	N·m
M_r	relative molecular mass, relative molar mass	1	
M	magnetization	$L^{-1}I$	A/m
N	neutron number: $A - Z$	1	
	number of particles	1	

		Dimension	SI Unit
N_E	density of states: $dN(E)/dE$	$L^{-5}M^{-1}T^2$	J^{-1}/m^3
N_ω	(spectral) density of vibrational modes	$L^{-3}T$	s/m^3
P	power	L^2MT^{-3}	W, J/s
	pressure	$L^{-1}MT^{-2}$	Pa, J/m^3
	probability density	L^{-3}	m^{-3}
$\boldsymbol{P}$	electric polarization: $\boldsymbol{D} - \epsilon_\circ \boldsymbol{P}$	$L^{-2}TI$	C/m^2
Q	quadrupole moment	L^2TI	$C{\cdot}m^2$
	quality factor	1	
	quantity of electricity, charge	TI	C
	quantity of heat	L^2MT^{-2}	J
	quantity of light	TJ	lm·s
	reaction energy, disintegration energy	L^2MT^{-2}	J, eV
R	radius, nuclear radius, range	L	m
	resistance, reluctance	$L^2MT^{-1}I^{-2}$	Ω
	thermal resistance	$L^{-2}m^{-1}T^3\Theta$	K/W
R_H	Hall coefficient	$M^3T^{-1}I^{-1}$	m^3/C
$\boldsymbol{R}$	lattice vector	L	m
S	area	L^2	m^2
	entropy	$L^2MT^{-2}\Theta^{-1}$	J/K
	spin quantum number	1	
	stopping power	LMT^{-2}	J/m, eV/m
S_a	atomic stopping power	L^4MT^{-2}	$J{\cdot}m^2$, $eV{\cdot}m^2$
S_{ab}	Seebeck coefficient	$L^2MT^{-3}I^{-1}\Theta^{-1}$	V/K
$\boldsymbol{s}$	Poynting vector	$L^{-1}MT^{-3}$	W/m^2
T	kinetic energy	L^2MT^{-2}	J
	period, periodic time	T	s
	torque, moment of a couple	L^2MT^{-2}	N·m
$T_{\frac{1}{2}}$	half-life	T	s
T_c	superconductor critical transition temperature	Θ	K
T_C	Curie temperature	Θ	K
T_N	Néel temperature	Θ	K
U	(electrical) potential difference	$L^2MT^{-3}I^{-1}$	V
	potential energy, thermodynamic energy	L^2MT^{-2}	J
U_m	magnetic potential difference	I	A
V	electric potential difference	$L^2MT^{-3}I^{-1}$	V
	potential energy	L^2MT^{-2}	J

		Dimension	SI Unit
	volume	L^3	m^3
W	energy	L^2MT^{-2}	J
	weight	LMT^{-2}	N
	work: $\int \boldsymbol{F} \cdot d\boldsymbol{s}$	L^2MT^{-2}	J
X	reactance	$L^2MT^{-1}I^{-2}$	Ω
	exposure (x- or γ-ray)	$M^{-1}TI$	C/kg
Y	admittance: $Y = 1/Z = G + jB$	$L^2MT^{-1}I^{-2}$	Ω
	Young's modulus	$L^{-1}MT^{-2}$	$Pa, N/m^2$
Z	atomic number	1	
	impedance: $R + jX$	$L^2MT^{-1}I^{-2}$	Ω
α	absorption factor, absorbance	1	
	angular acceleration	T^{-2}	s^{-2}
	annihilation operator	1	
	attenuation factor	L^{-1}	m^{-1}
	cubic expansion coefficient	Θ^{-1}	K^{-1}
	internal conversion coefficient	1	
	Madelung constant	1	
	plane angle	1	rad
	(electric) polarizability	$M^{-1}T^4I^2$	$C \cdot m^2/V$
	recombination coefficient	L^3T^{-1}	m^3/s
$\alpha^\dagger$	creation operator	1	
α_T	thermal diffusion factor	1	
β	annihilation operator	1	
	plane angle	1	rad
$\beta^\dagger$	creation operator	1	
γ	conductivity: $1/\rho$	$L^{-3}M^{-1}T^3I^2$	S/m
	growth rate	T^{-1}	s^{-1}
	gyromagnetic ratio: ω/B	$M^{-1}TI$	s^{-1}/T, C/kg
	plane angle	1	rad
	shear strain	1	
	surface tension	MT^{-2}	J/m^2, N/m
γ, Γ	Grüneisen parameter: $\alpha/(\kappa_T c_V \rho)$	1	
δ	damping coefficient	T^{-1}	s^{-1}
	loss angle: $\arctan(1/Q)$	1	rad
	thickness	L	m
ϵ	emissivity	1	
	linear strain	1	
	permittivity	$L^{-3}M^{-1}T^4I^2$	F/m
ϵ_{ab}	Seebeck coefficient	$L^2MT^{-3}I^{-1}\Theta^{-1}$	V/K
ϵ_{ij}	strain tensor	1	

		Dimension	SI Unit
ϵ_r	relative permittivity	1	
ϵ_F	Fermi energy	L^2Mt^{-2}	J, eV
η	viscosity	L^2MT^{-1}	Pa·s
θ	plane angle, scattering angle	1	
ϑ	Bragg angle, scattering angle	1	
	Celsius temperature	Θ	K, °C
κ	bulk modulus, compressibility	$LM^{-1}T^2$	M²/N, Pa⁻¹
	electrolytic conductivity	$L^{-3}M^{-1}T^3I^2$	S/m
	Landau–Ginzburg parameter	1	
λ	absolute activity: $\exp(\mu/kT)$	1	
	damping coefficient, disintegration constant, decay constant	T^{-1}	s⁻¹
	mean free path	L	m
	thermal conductivity	$LMT^{-3}\Theta^{-1}$	W/(m·K)
	wavelength	L	m
λ_C	Compton wavelength: h/mc	L	m
λ_L	London penetration depth	L	m
μ	chemical potential	$L^2MT^{-2}N^{-1}$	J/mol
	linear attenuation coefficient	L^{-1}	m⁻¹
	electric dipole moment	LTI	C·m
	magnetic dipole moment	L^2I	A·m²
	permeability	$LMT^{-2}I^{-2}$	N/A²
	Poisson ratio	1	
	reduced mass: $m_1m_2/(m_1+m_2)$	M	kg
	shear modulus	$L^{-1}MT^{-2}$	Pa, N/m²
μ_a	atomic attenuation coefficient	L^2	m²
μ_m	mass attenuation coefficient	L^2M^{-1}	m²/kg
μ_r	relative permeability: $\mu/\mu_\circ$	1	
ν	amount of substance	N	mol
	frequency	T^{-1}	s⁻¹
	kinematic viscosity: η/ρ	L^2T^{-1}	m²/s
ν_B	stoichiometric number of substance B	1	
ξ	coherence length	L	m
	particle displacement	L	m
ρ	charge density	$L^{-3}TI$	C/m³
	reflection coefficient	1	
	resistivity	$L^3MT^{-3}I^{-1}$	Ω·m
	(mass) density	$L^{-3}M$	kg/m³
ρ_R	residual resistivity	$L^3MT^{-3}I^{-1}$	Ω·m
σ	conductivity: $1/\rho$	$L^{-3}M^{-1}T^3I^2$	S/m

		Dimension	SI Unit
	cross section	L^2	m^2
	normal stress	$L^{-1}MT^{-2}$	Pa, N/m^2
	short-range order parameter	1	
	surface charge density	$L^{-2}TI$	C/m^2
	surface tension	$L^{-1}MT^{-2}$	N/m, J/m^3
	wave number	L^{-1}	m^{-1}
σ	wave vector	L^{-1}	m^{-1}
τ	mean life, relaxation time	T	s
	shear stress	$L-1MT^{-2}$	Pa, N/m^2
	transmission coefficient	1	
τ_{ij}	stress tensor	1	
τ_m	mean life	T	s
$\tau_{1/2}$	half-life	T	s
ϕ	electric potential	$L^2MT^{-3}I^{-1}$	V
	osmotic coefficient	1	
	(particle) fluence rate, flux density	$L^{-2}T$	m^{-2}/s
	phase difference	1	rad
	plane angle	1	
	volume fraction	1	
χ	(magnetic) susceptibility	1	
χ_e	electric susceptibility	1	
χ_m	magnetic susceptibility	1	
ψ	radiant energy fluence rate	MT^{-3}	W/m^2
ω	angular frequency: $2\pi f$	T^{-1}	s
	solid angle	1	sr
ω_D	Debye angular frequency	T^{-1}	s^{-1}
ω_L	Larmor circular frequency	T^{-1}	s^{-1}
Γ	level width	L^2MT^{-2}	J, eV
Δ	superconductor energy gap	L^2MT^{-2}	J, eV
Θ_{rot}	characteristic rotational temperature: $h^2/8\pi^2kI$	Θ	K
Θ_{vib}	characteristic vibrational temperature: $h\nu/k$	Θ	K
Θ_D	Debye temperature: $h\nu_D/k$	Θ	K
Θ_E	Einstein temperature: $h\nu_E/k$	Θ	K
Θ_W	Weiss temperature	Θ	K
Λ	logarithmic decrement	1	Np
	mean free path of phonons	L	m
Π	osmotic pressure	$L^{-1}MT^{-2}$	Pa, N/m^2
Π_{ab}	Peltier coefficient	$L^2MT^{-3}I^{-1}$	V

		Dimension	SI Unit
Σ	macroscopic cross section: $n\sigma$	L^{-1}	m^{-1}
Φ	luminous flux	J	lm
	magnetic flux	$L^2MT^{-2}I^{-2}$	Wb, V·s
Ψ	electric flux	TI	C
	solid angle	1	sr

3.2   Quantum mechanics

3.2.1   Matrix element and operator symbols

A_{ij}	matrix element:	$\int \phi_i^* (A\phi_j)\, d\tau$
$A^\dagger$	Hermitian conjugate of A:	$(A^\dagger)_{ij} = A_{ji}^*$
$\langle A \rangle$	expectation value of A:	$\mathrm{Tr}\,(A)$
$[A, B]$	commutator of A and B:	$AB - BA$
$[A, B]_-$	commutator of A and B:	$AB - BA$
$[A, B]_+$	anticommutator of A and B:	$AB + BA$
$\langle ...|$	Dirac bra vector	
$|...\rangle$	Dirac ket vector	

Pauli matrices: σ, I

$$\sigma_x = \begin{pmatrix} 0 & 1 \\ 1 & 0 \end{pmatrix}, \quad \sigma_y = \begin{pmatrix} 0 & -i \\ i & 0 \end{pmatrix}, \quad \sigma_z = \begin{pmatrix} 1 & 0 \\ 0 & -1 \end{pmatrix}, \quad I = \begin{pmatrix} 1 & 0 \\ 0 & 1 \end{pmatrix}$$

Dirac matrices: α, β

$$\alpha_x = \begin{pmatrix} 0 & \sigma_x \\ \sigma_x & 0 \end{pmatrix}, \quad \alpha_y = \begin{pmatrix} 0 & \sigma_y \\ \sigma_y & 0 \end{pmatrix}, \quad \alpha_z = \begin{pmatrix} 0 & \sigma_z \\ \sigma_z & 0 \end{pmatrix}, \quad \beta = \begin{pmatrix} I & 0 \\ 0 & -I \end{pmatrix}$$

3.3   Crystallography

Fundamental translation vectors for the crystal lattice:

$$\boldsymbol{R} = n_1 \boldsymbol{a}_1 + n_2 \boldsymbol{a}_2 + n_3 \boldsymbol{a}_3 \quad (n_1,\ n_2,\ n_3,\ \text{integers}).$$

Fundamental translation vectors for the reciprocal lattice:

$$b_1, b_2, b_3 \qquad a^*, b^*, c^*.$$

In crystallography $\boldsymbol{a}_i \cdot \boldsymbol{b}_k = \delta_{ik}$; in solid-state physics $\boldsymbol{a}_i \cdot \boldsymbol{b}_k = 2\pi\delta_{ik}$.

Indices of the Bragg reflection from the lattice are denoted by h, k, l:
(h, k, l) single plane or set of parallel planes in a lattice
$\{h, k, l\}$ set of all symmetry-equivalent lattice planes

$[u, v, w]$ indices of a lattice direction

$\langle u, v, w \rangle$ set of all symmetry-equivalent lattice directions.

When the letter symbols are replaced by numbers, it is customary to omit the commas and to represent negative numbers by a bar or overline above the number; e.g., $(12\overline{2})$ denotes the planes $h = 1$, $k = 2$, $l = -2$. $d_{hkl} = d_{12\overline{2}}$.

References

[1] *Symbols, Units, Nomenclature and Fundamental Constants in Physics*, IUPAP-25 (1987), E. R. Cohen and P. Giacomo [Physica **146A**, (1987)].

[2] International Organization for Standardization, *Quantities and Units*, *(ISO Standards Handbook*, Geneva, 1993).

[3] *Quantities, Units and Symbols in Physical Chemistry*, edited by I. M. Mills and T. Cvitaš (Blackwell Scientific, Oxford, 1993).

4

PRECISE PHYSICAL CONSTANTS

4.1 Standards and conversion factors

4.1.1 Internationally adopted electrical standards

The Josephson Effect and the quantized Hall Effect provide natural standards for the volt and the ohm respectively.

Josephson Effect: The dc current passing through a Josephson junction irradiated with microwave radiation of frequency f exhibits discontinuities at quantized potential steps $U(n) = nU_{\mathrm{J}} = nf/K_{\mathrm{J}}$, where n is an integer and K_{J} is a constant independent of the superconductors that make up the junction. Since 1972, the U.S. representation of the volt had been defined by the value

$$K_{\mathrm{J-NBS72}} = 483\,593.420\,\mathrm{GHz/V}$$

while most other national laboratories that used a Josephson standard followed the BIPM recommendation

$$K_{\mathrm{J-72}} = 483\,594\,\mathrm{GHz/V}.$$

Subsequently, more accurate data led to a revised value that, since January 1990, has been adopted as the standard by all national laboratories:

$$K_{\mathrm{J-90}} \equiv 483\,597.9\,\mathrm{GHz/V}.$$

An ideal realization of the volt in terms of the Josephson effect using this value for K_{J} will reproduce the SI volt with a relative uncertainty of 4 part in 10^7 (one standard deviation).

Quantized Hall Effect: The ratio of the transverse Hall potential to the longitudinal current in a two-dimensional electron gas at a semiconductor heterojunction at high magnetic field and sufficiently low temperature (the von Klitzing effect) is given by $V/I = R_K/i$ when the longitudinal potential approaches zero, where the integer i counts the number of Landau levels below the Fermi level. The value

$$R_{K-90} = 25\,812.807\,\Omega$$

has been adopted as an international standard of resistance that reproduces the SI ohm with an overall uncertainty of 2 parts in 10^7 (one standard deviation).

4.1.2   X-ray standards

Table 4.1 X-ray wavelength standards.   The digits in parentheses represent the one-standard-deviation uncertainty in the final digits of the value.

Quantity	Symbol	Value
Cu x-unit: [$\lambda(CuK\alpha_1) \equiv 1537.400\,xu$]	$xu(CuK\alpha_1)$	$1.002\,077\,89(70)\times10^{-13}$ m
Mo x-unit: [$\lambda(MoK\alpha_1) \equiv 707.831\,xu$]	$xu(MoK\alpha_1)$	$1.002\,099\,38(45)\times10^{-13}$ m
Å*: [$\lambda(WK\alpha_1) \equiv 0.209\,0100\,\text{Å}^*$]	Å*	$1.000\,014\,81(92)\times10^{-10}$ m
Lattice spacing of Si [a] (in vacuum, 22.5 °C)	a	$0.543\,101\,96(11)$ nm
$d_{220} = a/\sqrt{8}$	d_{220}	$0.192\,015\,540(40)$ nm
Molar volume of Si, $M(Si)/\rho(Si) = N_A a^3/8$	$V_m(Si)$	$12.058\,8179(89)$ cm^3/mol

[a]   The lattice spacing of single-crystal Si can vary by parts in 10^7 depending on the preparation process, which can also produce distortions from exact cubic symmetry of the same order.

4.2 Unit systems

Table 4.2 Atomic units. The atomic constants, used for calculations in atomic and molecular physics are equivalent to setting $m_e = 1$, $e = 1$, $\hbar = 1$, $4\pi\epsilon_0 = 1$. In this system the speed of light is $c = 1/\alpha \approx 137$.

Quantity	Identification	Symbol	Value
Length	Bohr radius	a_0	$0.529\,177\,249(24)\times 10^{-10}$ m
Mass	Electron mass	m_e	$0.910\,938\,97(54)\times 10^{-30}$ kg
Time	$a_0/\alpha c$		$2.418\,884\,337(3)\times 10^{-17}$ s
Velocity, speed	$c = 1/\alpha$		$2.187\,691\,42(10)\times 10^{6}$ m/s
Angular momentum, action		$\hbar$	$1.054\,572\,66(63)\times 10^{-34}$ J s
Energy (hartree)	$2R_\infty hc$	E_h	$4.359\,7482(26)\times 10^{-18}$ J
Electric charge		e	$1.602\,177\,33(49)\times 10^{-19}$ C
Electric current	$eE_h/\hbar$		$6.623\,6211(20)\times 10^{-3}$ A
Potential	E_h/e		$27.211\,3962(82)$ V

Table 4.3 Energy equivalents.

Quantity	Symbol	Value
Atomic mass unit	u	$931.494\,32(28)$ MeV/c^2
Electron mass	m_e	$0.510\,999\,06(15)$ MeV/c^2
Muon mass	m_μ	$105.658\,389(34)$ MeV/c^2
Proton mass	m_p	$938.272\,31(28)$ MeV/c^2
Neutron mass	m_n	$939.565\,63(28)$ MeV/c^2
Electron volt	1 eV	$1.602\,177\,33(49)\times 10^{-19}$ J
	1 eV/hc	$8065.5410(24)$ cm^{-1}
	1 eV/h	$241.798\,836(73)$ THz
Electron volt per particle	1 eV/k	$11\,604.45(10)$ K
Planck constant	$\hbar$	$6.582\,1220(20)\times 10^{-22}$ MeV s
	$\hbar c$	$197.327\,053(60)$ MeV fm
	$(\hbar c)^2$	$0.389\,379\,66(23)$ GeV^2 mb
Rydberg constant	$R_\infty hc$	$13.605\,6981(41)$ eV
Voltage-wavelength product	$V\cdot\lambda$	$12\,398.4244(37)$ eV Å

Table 4.4 Quantum-gravity (Planck) units. These units are defined by setting $\hbar$, G, and c equal to unity. The uncertainties in the numerical values are dominated by the uncertainty of $\sqrt{G}$, 64 ppm.

Quantity	Identification	Symbol	Value
Mass	$(\hbar c/G)^{\frac{1}{2}}$	m_{Pl}	$2.176\,71(14)\times10^{-8}$ kg
Length	$\hbar/m_{\mathrm{Pl}}c = (\hbar G/c^3)^{\frac{1}{2}}$	l_{Pl}	$1.616\,05(10)\times10^{-35}$ m
Time	$l_{\mathrm{Pl}}/c = (\hbar G/c^5)^{\frac{1}{2}}$	t_{Pl}	$5.390\,56(34)\times10^{-44}$ s
Energy	$m_{\mathrm{Pl}}c^2 = (\hbar c^5/G)^{\frac{1}{2}}$		$1.956\,33(12)\times10^{9}$ J
			$1.221\,05(8)\times10^{19}$ GeV

Table 4.5 Astronomical units. The astronomical system of units is defined by the Gaussian constant of proper motion:

$$k \equiv 0.017\,202\,098\,95 \,\mathrm{rad/day} = 1.990\,983\,674\,77\times10^{-7}\,\mathrm{s^{-1}},$$

such that the gravitational constant is $G = k^2\,\mathrm{AU}^3/M_\odot$.

Quantity	Identification	Symbol	Value
Mass	Solar mass	$M_\odot$	1.9891×10^{30} kg
Length	AU	$\tau_A c$	$1.495\,9787\times10^{11}$ m
		τ_A	$499.004\,782$ s
Time	day		$86\,400$ s
Constant of gravitation	$k^2\,\mathrm{AU}^3$	$GM_\odot$	$1.327\,231\,152\times10^{20}$ m³/s²

4.3 Physical constants

Table 4.6 Geodetic constants (IUGG 1980).

Quantity	Symbol	Value
Earth equatorial radius	a_e	$6\,378\,137$ m
Geocentric gravitational constant	GM	$3\,986\,005\times10^{8}$ m³ s⁻²
Angular velocity	ω	$7\,292\,115\times10^{-11}$ s⁻¹
Dynamic form factor	J_2	$0.001\,082\,63$
Derived values:		
Earth polar radius	a_p	$6\,356\,752.3141$ m
flattening	f	$1/298.257\,222\,101$
1° of latitude		$(110\,575 + 1\,110\sin^2\phi)$ m
1° of longitude		$(111\,320 + 373\sin^2\phi)\cos\phi$ m
Normal gravity	$g(\phi) = 9.780\,326\,7715\dfrac{1 + 0.001\,931\,851\,353\sin^2\phi}{(1 - 0.006\,694\,380\,023\sin^2\phi)^{1/2}}$ m/s²	
Geodetic latitude − geocentric latitude	$\phi - \phi' = 692.74''\sin 2\phi - 1.16''\sin 4\phi$	

Table 4.7 Astronomical constants (IAU, 1976). The figures given in brackets are updated values used in the 1992 ephemerides.

Quantity	Symbol	Value
Gaussian constant	k	0.017 202 098 95 rad/d
	k'	$1.990\,983\,674\,77\times10^{-7}$ rad/s
Ephemeris (tropical) year		
(1900.0)	yr	31 556 925.9747 s
		365.242 198 78 d
		365 d 5 h 48 m 45.9747 s
(2000.0) [a]	yr	31 556 925.187 47 s
		365.242 189 6698 d
		365 d 5 h 48 m 45.187 47 s
Astronomical unit	AU	499.004 782 s
		[499.004 7837]
		$1.495\,9787\times10^{11}$ m
		[1.495 978 7066]
Mean distance, Earth to Sun		1.000 001 0178 AU
		[1.000 001 057 266 65]
		$1.495\,980\,224\times10^{11}$ m
		[1.495 980 2882]
Mean orbital speed		29.7859 m/s
		[29.784 766 966]
Mean centripetal acceleration		0.005 94 m/s^2
		[0.005 930 113 4387]
Earth gravitational constant	$GM_\oplus$	$3.986\,005\times10^{14}$ m^3/s^2
		[3.986 004 48]
Dynamic form factor	J_2	0.001 082 63
Mass ratios		
Sun/Earth	$M_\odot/M_\oplus$	332 946.0
		[332 946.038]
Earth/Moon	$M_\oplus/M_m$	81.300 68
		[81.300 59]
Sun/(Earth + Moon)	$M_\odot/(M_\oplus+M_m)$	328 900.5
		[328 900.55]
Radius of the Sun		6.96×10^8 m
Surface gravity		274 m/s^2 = 27.9 g$_o$
Schwarzschild radius	$2GM_\odot/c^2$	2953.487 631 m
Motion of solar system	speed:	0.0112 AU/d = 19 400 m/s
relative to nearby stars	apex:	$\alpha = 271°$ $\delta = +30°$
Mass of the Sun	$M_\odot$	1.9891×10^{30} kg
Mean density	ρ_S	1410 kg/m^3
of the Earth	$M_\oplus$	5.9742×10^{24} kg
Mean density	ρ_E	5520 kg/m^3
of the Moon	M_m	$7.370\,37\times10^{22}$ kg
Sidereal year (1990)		365d6h9m10s
Julian year (365.25 d)		31 557 600 s
Gregorian year (365.2425 d = 365d 5h 49m 12s)		31 556 952 s
Light-year[b]	ly	$9.460\,730\times10^{15}$ m
Parsec [AU/sin(1″)]	pc	$3.085\,673\times10^{16}$ m
		3.261 564 ly
Julian day[c]	JD	JD2450000 = 9 Oct 1995
	JD	JD2451545 = 1 Jan 2000

[a] Ref. 3. [b] The "year" here is the Julian year, 36525 days per Julian century. [c] The Julian day runs from noon (Universal Time) on the given date to noon of the following day.

4.4 Fundamental constants

The numerical values of the fundamental physical quantities listed in these tables are based on the 1986 CODATA recommendations.[4] (Newer information and possible changes in these values are discussed by B. N. Taylor and E. R. Cohen,[5] and in the annual listing of the fundamental physical constants in the August issues of *Physics Today*. A formal revised recommendation is expected to be available in 1996.)

In the following tables, the digits in parentheses give the uncertainty of the final digits of the entry. Thus $6.672\,59(85)$ indicates that the uncertainty in the value $6.672\,59$ is $0.000\,85$. This uncertainty (expressed as a relative uncertainty in the last column) is intended to represent one standard deviation. Since the uncertainties in these tables are correlated, the full covariance matrix of the least-squares solution must be used to evaluate the uncertainty associated with quantities calculated from these entries.

Table 4.8 Universal constants.

Quantity	Symbol	Value	Relative uncertainty (ppm)
Speed of light in vacuum	c	$299\,792\,458\,\mathrm{m\,s^{-1}}$	(exact)
Permeability of vacuum	μ_0	$4\pi \times 10^{-7}\,\mathrm{N\,A^{-2}}$	
		$1.256\,637\,0614\ldots\times10^{-6}\,\mathrm{N\,A^{-2}}$	(exact)
Permittivity of vacuum, $1/\mu_0 c^2$	ϵ_0	$8.854\,187\,817\ldots\times10^{-12}\,\mathrm{F\,m^{-1}}$	(exact)
Newtonian constant of gravitation	G	$6.672\,59(85)\times10^{-11}\,\mathrm{m^3\,kg^{-1}\,s^{-2}}$	128
Planck constant	h	$6.626\,0755(40)\times10^{-34}\,\mathrm{J\,s}$	0.60
in electron volts, $h/\{e\}$		$4.135\,6692(12)\times10^{-15}\,\mathrm{eV\,s}$	0.30
$h/2\pi$	$\hbar$	$1.054\,572\,66(63)\times10^{-34}\,\mathrm{J\,s}$	0.60
in electron volts, $\hbar/\{e\}$		$6.582\,1220(20)\times10^{-16}\,\mathrm{eV\,s}$	0.30
Fine-structure constant, $\dfrac{\mu_0 c}{2}\cdot\dfrac{e^2}{h} = \dfrac{e^2}{4\pi\epsilon_0 \hbar c}$	α	$7.297\,353\,08(33)\times10^{-3}$	0.045
Inverse fine-structure constant	α^{-1}	$137.035\,9895(61)$	0.045

Table 4.9 Atomic constants.

Quantity	Symbol	Value	Relative uncertainty (ppm)
Elementary charge	e	$1.602\,177\,33(49)\times10^{-19}$ C	0.30
	e/h	$2.417\,988\,36(72)\times10^{14}$ A J^{-1}	0.30
Magnetic flux quantum, $h/2e$	$\Phi_\circ$	$2.067\,834\,61(61)\times10^{-15}$ Wb	0.30
Josephson frequency–voltage ratio	$2e/h$	$4.835\,9767(14)\times10^{14}$ Hz V^{-1}	0.30
Quantized Hall conductance	e^2/h	$3.874\,046\,14(17)\times10^{-5}$ S	0.045
Quantized Hall resistance, $h/e^2 = \mu_\circ c\big/2\alpha$	R_H	$25\,812.8056(12)$ Ω	0.045
Bohr magneton, $e\hbar/2m_e$	μ_B	$9.274\,0154(31)\times10^{-24}$ J T^{-1}	0.34
in electron volts, $\mu_B/\{e\}$		$5.788\,382\,63(52)\times10^{-5}$ eV T^{-1}	0.089
in hertz, μ_B/h		$1.399\,624\,18(42)\times10^{10}$ Hz T^{-1}	0.30
in wave numbers, μ_B/hc		$46.686\,437(14)$ m^{-1} T^{-1}	0.30
in kelvins, μ_B/k		$0.671\,7099(57)$ K T^{-1}	8.5
Nuclear magneton, $e\hbar/2m_p$	μ_N	$5.050\,7866(17)\times10^{-27}$ J T^{-1}	0.34
in electron volts, $\mu_N/\{e\}$		$3.152\,451\,66(28)\times10^{-8}$ eV T^{-1}	0.089
in hertz, μ_N/h		$7.622\,5914(23)$ MHz T^{-1}	0.30
in wave numbers, μ_N/hc		$2.542\,622\,81(77)\times10^{-2}$ m^{-1} T^{-1}	0.30
in kelvins, μ_N/k		$3.658\,246(31)\times10^{-4}$ K T^{-1}	8.5
Rydberg constant, $m_e c\alpha^2\big/2h$	R_∞	$10\,973\,731.534(13)$ m^{-1}	0.0012
in hertz, $R_\infty c$		$3.289\,841\,9499(39)\times10^{15}$ Hz	0.0012
in joules, $R_\infty hc$		$2.179\,8741(13)\times10^{-18}$ J	0.60
in eV, $R_\infty hc/\{e\}$		$13.605\,6981(40)$ eV	0.30
Bohr radius, $\alpha/4\pi R_\infty$	$a_\circ$	$0.529\,177\,249(24)\times10^{-10}$ m	0.045
Hartree energy, $2R_\infty hc = e^2/4\pi\epsilon_\circ a_\circ$	E_h	$4.359\,7482(26)\times10^{-18}$ J	0.60
in eV, $E_h/\{e\}$		$27.211\,3961(81)$ eV	0.30
Quantum of circulation	$h/2m_e$	$3.636\,948\,07(33)\times10^{-4}$ m^2 s^{-1}	0.089
	h/m_e	$7.273\,896\,14(65)\times10^{-4}$ m^2 s^{-1}	0.089

ELECTRON

Quantity	Symbol	Value	Relative uncertainty (ppm)
Electron mass	m_e	$9.109\,3897(54)\times10^{-31}$ kg	0.59
		$5.485\,799\,03(13)\times10^{-4}$ u	0.023
		$0.510\,999\,06(15)$ MeV/c^2	0.30
Mass ratios:			
electron–muon	m_e/m_μ	$0.004\,836\,332\,18(71)$	0.15
electron–proton	m_e/m_p	$5.446\,170\,13(11)\times10^{-4}$	0.020
electron–deuteron	m_e/m_d	$2.724\,437\,07(6)\times10^{-4}$	0.020
electron–α-particle	m_e/m_α	$1.370\,933\,54(3)\times10^{-4}$	0.021
Specific charge	$-e/m_e$	$-1.758\,819\,62(53)\times10^{11}$ C kg^{-1}	0.30
Molar mass	M_e	$5.485\,799\,03(13)\times10^{-7}$ kg/mol	0.023
Compton wavelength, $h/m_e c$	λ_C	$2.426\,310\,58(22)\times10^{-12}$ m	0.089
$\lambda_C/2\pi = \alpha^2\big/4\pi R_\infty$	$\lambdabar_C$	$3.86\,159\,323(35)\times10^{-13}$ m	0.089

Table 4.9 *Continued.*

Quantity	Symbol	Value	Relative uncertainty (ppm)
Classical electron radius, $\alpha^2 a_\circ$	r_e	$2.817\,940\,92(38) \times 10^{-15}$ m	0.13
Thomson cross section, $(8\pi/3)r_e^2$	σ_e	$0.665\,246\,16(18) \times 10^{-28}$ m^2	0.27
Magnetic moment	μ_e	$928.477\,01(31) \times 10^{-26}$ J T^{-1}	0.34
in Bohr magnetons	μ_e/μ_B	$1.001\,159\,653\,193(10)$	1×10^{-5}
in nuclear magnetons	μ_e/μ_N	$1838.282\,000(37)$	0.020
Magnetic moment anomaly, $\mu_e/\mu_B - 1$	a_e	$1.159\,653\,193(10) \times 10^{-3}$	0.0086
g-factor, $2(1 + a_e)$	g_e	$2.002\,319\,304\,386(20)$	1×10^{-5}
Magnetic moment ratios:			
electron–muon	μ_e/μ_μ	$206.766967(30)$	0.15
electron–proton	μ_e/μ_p	$658.210\,6881(66)$	0.010

MUON

Quantity	Symbol	Value	Relative uncertainty (ppm)
Muon mass	m_μ	$1.883\,5327(11) \times 10^{-28}$ kg	0.61
		$0.113\,428\,913(17)$ u	0.15
		$105.658\,389(34)$ MeV/c^2	0.32
Muon–electron mass ratio	m_μ/m_e	$206.768\,262(30)$	0.15
Molar mass	M_μ	$1.134\,289\,13(17) \times 10^{-4}$ kg/mol	0.15
Magnetic moment	μ_μ	$4.490\,4514(15) \times 10^{-26}$ J T^{-1}	0.33
in Bohr magnetons,	μ_μ/μ_B	$4.841\,970\,97(71) \times 10^{-3}$	0.15
in nuclear magnetons,	μ_μ/μ_N	$8.890\,5981(13)$	0.15
Magnetic moment anomaly, $\mu_\mu/(e\hbar/2m_\mu) - 1$	a_μ	$1.165\,9230(84) \times 10^{-3}$	7.2
g-factor, $2(1 + a_\mu)$	g_μ	$2.002\,331\,846(17)$	0.0084
Muon–proton magnetic moment ratio	μ_μ/μ_p	$3.183\,345\,47(47)$	0.15

PROTON

Quantity	Symbol	Value	Relative uncertainty (ppm)
Proton mass	m_p	$1.672\,6231(10) \times 10^{-27}$ kg	0.59
		$1.007\,276\,470(12)$ u	0.012
		$938.272\,31(28)$ MeV/c^2	0.30
Mass ratios:			
proton–electron	m_p/m_e	$1836.152\,701(37)$	0.020
proton–muon	m_p/m_μ	$8.880\,2444(13)$	0.15
Specific charge	e/m_p	$9.578\,8309(29) \times 10^7 \cdot$C kg^{-1}	0.30
Molar mass	M_p	$1.007\,276\,470(12) \times 10^{-3}$ kg/mol	0.012
Compton wavelength, $h/m_p c$	$\lambda_{C,p}$	$1.321\,410\,02(12) \times 10^{-15}$ m	0.089
$\lambda_{C,p}/2\pi$	$\lambda_{C,p}$	$2.103\,089\,37(19) \times 10^{-16}$ m	0.089
Magnetic moment	μ_p	$1.410\,607\,61(47) \times 10^{-26}$ J T^{-1}	0.34
in Bohr magnetons	μ_p/μ_B	$1.521\,032\,202(15) \times 10^{-3}$	0.010
in nuclear magnetons	μ_p/μ_N	$2.792\,847\,386(63)$	0.023

Table 4.9 *Continued.*

Quantity	Symbol	Value	Relative uncertainty (ppm)
Shielded proton moment			
(water, spherical, 25 °C)	μ'_p	$1.410\,571\,38(47) \times 10^{-26}$ J T^{-1}	0.34
in Bohr magnetons	μ'_p/μ_B	$1.520\,993\,129(17) \times 10^{-3}$	0.011
in nuclear magnetons	μ'_p/μ_N	$2.792\,775\,642(64)$	0.023
Diamagnetic shielding			
correction, $\mu'_p/\mu_p - 1$	σ_{H_2O}	$-25.689(15) \times 10^{-6}$	580.0
Proton gyromagnetic ratio	γ_p	$26\,752.2128(81) \times 10^4 \cdot$s^{-1} T^{-1}	0.30
	$\gamma_p/2\pi$	$42.577\,469(13)$ MHz T^{-1}	0.30
uncorrected	γ'_p	$26\,751.5255(81) \times 10^4 \cdot$s^{-1} T^{-1}	0.30
	$\gamma'_p/2\pi$	$42.576\,375(13)$ MHz T^{-1}	0.30

NEUTRON

Quantity	Symbol	Value	Relative uncertainty (ppm)
Neutron mass	m_n	$1.674\,9286(10) - 27$ kg	0.59
		$1.008\,664\,904(14)$ u	0.014
		$939.565\,63(28)$ MeV/c^2	0.30
Mass ratios:			
neutron–electron	m_n/m_e	$1838.683\,662(40)$	0.022
neutron–proton	m_n/m_p	$1.001\,378\,404(9)$	0.009
Molar mass	M_n	$1.008\,664\,904(14) \times 10^{-3}$ kg/mol	0.014
Compton wavelength, $h/m_n c$			
	$\lambda_{C,n}$	$1.319\,591\,10(12) \times 10^{-15}$ m	0.089
$\lambda_{C,n}/2\pi$	$\lambda_{C,n}$	$2.100\,194\,45(19) \times 10^{-16}$ m	0.089
Magnetic moment	μ_n	$-0.966\,237\,07(40) \times 10^{-26}$ J T^{-1}	0.41
in Bohr magnetons	μ_n/μ_B	$-1.041\,875\,63(25) \times 10^{-3}$	0.24
in nuclear magnetons	μ_n/μ_N	$-1.913\,042\,75(45)$	0.24
Magnetic moment ratios:			
neutron–electron	μ_n/μ_e	$1.040\,668\,82(25) \times 10^{-3}$	0.24
neutron–proton	μ_n/μ_p	$-0.684\,979\,34(16)$	0.24

DEUTERON

Quantity	Symbol	Value	Relative uncertainty (ppm)
Deuteron mass	m_d	$3.343\,5860(20) \times 10^{-27}$ kg	0.59
		$2.013\,553\,214(24)$ u	0.012
		$1875.613\,39(57)$ MeV/c^2	0.30
Mass ratios:			
deuteron–electron	m_d/m_e	$3670.483\,014(75)$	0.020
deuteron–proton	m_d/m_p	$1.999\,007\,496(6)$	0.003
Molar mass	M_d	$2.013\,553\,214(24) \times 10^{-3}$ kg/mol	0.012
Magnetic moment	μ_d	$0.433\,073\,75(15) \times 10^{-26}$ J T^{-1}	0.34
in Bohr magnetons,	μ_d/μ_B	$0.466\,975\,4479(91) \times 10^{-3}$	0.019
in nuclear magnetons,	μ_d/μ_N	$0.857\,438\,230(24)$	0.028
Magnetic moment ratios:			
deuteron–electron	μ_d/μ_e	$-0.466\,434\,5460(91) \times 10^{-3}$	0.019
deuteron–proton	μ_d/μ_p	$0.307\,012\,2035(51)$	0.017

Table 4.10 Physico-chemical constants.

Quantity	Symbol	Value	Relative uncertainty (ppm)
Avogadro constant	N_A, L	$6.022\,1367(36) \times 10^{23}\,\text{mol}^{-1}$	0.59
Atomic mass constant, $m_u = \frac{1}{12}m(^{12}\text{C})$	m_u	$1.660\,5402(10) \times 10^{-27}\,\text{kg}$	0.59
		$931.494\,32(28)\,\text{MeV}/c^2$	0.30
Faraday constant	F	$96\,485.309(29)\,\text{C}\,\text{mol}^{-1}$	0.30
Molar Planck constant	$N_A h$	$3.990\,313\,23(36) \times 10^{-10}\,\text{J}\,\text{s}\,\text{mol}^{-1}$	0.089
	$N_A hc$	$0.119\,626\,58(11)\,\text{J}\,\text{m}\,\text{mol}^{-1}$	0.089
Molar gas constant	R	$8.314\,510(70)\,\text{J}\,\text{mol}^{-1}\,\text{K}^{-1}$	8.4
Boltzmann constant, R/N_A	k	$1.380\,658(12) \times 10^{-23}\,\text{J}\,\text{K}^{-1}$	8.5
in electron volts, $k/\{e\}$		$8.617\,385(73) \times 10^{-5}\,\text{eV}\,\text{K}^{-1}$	8.4
in hertz, k/h		$2.083\,674(18) \times 10^{10}\,\text{Hz}\,\text{K}^{-1}$	8.4
in wave numbers, k/hc		$69.503\,87(59)\,\text{m}^{-1}\,\text{K}^{-1}$	8.4
Molar volume (ideal gas), RT_0/p_0, $T_0 = 273.15\,\text{K}$			
$p_0 = 101\,325\,\text{Pa}$	V_m	$22\,414.10(19)\,\text{cm}^3/\text{mol}$	8.4
Loschmidt constant, N_A/V_m	n_0	$2.686\,763(23) \times 10^{25}\,\text{m}^{-3}$	8.5
$p_0 = 100\,\text{kPa}$	V'_m	$22\,711.08(19)\,\text{cm}^3/\text{mol}$	8.4
Stefan-Boltzmann constant	σ	$5.670\,51(19) \times 10^{-8}\,\text{W}\,\text{m}^{-2}\,\text{K}^{-4}$	34
First radiation constant, $2\pi hc^2$	c_1	$3.741\,7749(22) \times 10^{-16}\,\text{W}\,\text{m}^2$	0.60
Second radiation constant, hc/k	c_2	$0.014\,387\,69(12)\,\text{m}\,\text{K}$	8.4
Wien displacement law constant, $b = \lambda_{\max}T = c_2/4.965\,114$	b	$2.897\,756(24) \times 10^{-3}\,\text{m}\,\text{K}$	8.4
Absolute entropy constant [a] (Sackur-Tetrode constant), $T_1 = 1\,\text{K}$, $p_0 = 100\,000\,\text{Pa}$	S_0/R	$-1.151\,693(21)$	18
$p_0 = 101\,325\,\text{Pa}$		$-1.164\,856(21)$	18

[a] The entropy of an ideal monatomic gas with atomic weight A is given by

$$S = \frac{5}{2}R + R\ln\left[\frac{kT}{p}\left(\frac{Am_u kT}{2\pi\hbar^2}\right)^{3/2}\right] = S_0 - R\ln\left(\frac{p}{p_0}\right) + \frac{3}{2}R\ln A + \frac{5}{2}R\ln\left(\frac{T}{T_1}\right)$$

where M° is the "standard molar mass," $M^\circ = 0.001\,\text{kg/mol}$, and

$$S_0/R = \frac{5}{2} + \ln\left[\frac{kT_1(2\pi M^\circ RT_1)^{3/2}}{p_0(N_A h)^3}\right].$$

Table 4.11 Covariance matrix for 1986 adjustment. This matrix gives the variances and covariances (relative uncertainty in (parts in 10^8)^2) for α^{-1}, h, μ_μ/μ_P, K_V, $K_V = (483\,594\,\text{GHz/V})h/2e$, and $K_\Omega = \Omega_{\text{B185}}/\Omega$, where Ω_{B185} is the BIPM representation of the ohm as of 1 January 1985. Because the covariance matrix is symmetric, the correlation coefficients are given (in italics) below the diagonal.

	α^{-1}	h	μ_μ/μ_P	K_V	K_Ω
α^{-1}	20	−41	33	890	9.25
h	*−0.154*	3582	−67	1770	−7.44
μ_μ/μ_P	*0.498*	*−0.077*	215	215	15.13
K_V	*−0.080*	*0.997*	*−0.040*	880	0.90
K_Ω	*−0.416*	*−0.025*	*0.207*	*0.006*	24.77

Table 4.12 Energy conversion factors. The entries in this table are given to five significant figures, and may be considered to be exact or be computed from the data given. More precise conversions can be found elsewhere in the tables in this chapter or can be computed from the data given there. Care should be taken, when the uncertainty of a computed conversion factor is evaluated, to include the contribution from correlations in the data if they are significant. To use this table note that the unit at the top of a column applies to all values beneath it; all entries on the same line are equal.

	J	kg	m^{-1}	Hz	K	eV	u	hartree
1 J =	1	$\{1/c^2\}$ 1.1126×10^{-17}	$\{1/hc\}$ 5.0341×10^{24}	$\{1/h\}$ 1.5092×10^{33}	$\{1/k\}$ 7.2430×10^{22}	$\{1/e\}$ 6.2415×10^{18}	$\{1/\{m_u c^2\}\}$ 6.7005×10^{9}	$\{c/2R_\infty h\}$ 2.2937×10^{17}
1 kg =	$\{c^2\}$ 8.9876×10^{16}	1	$\{c/h\}$ 4.5244×10^{41}	$\{c^2/h\}$ 1.3564×10^{50}	$\{c^2/k\}$ 6.5096×10^{39}	$\{c^2/e\}$ 5.6096×10^{35}	$\{1/m_u\}$ 6.0221×10^{26}	$\{c^2/2R_\infty h\}$ 2.0615×10^{34}
1 m^{-1} =	$\{hc\}$ 1.9864×10^{-25}	$\{h/c\}$ 2.2102×10^{-42}	1	$\{c\}$ 2.9979×10^{8}	$\{hc/k\}$ 0.01439	$\{hc/e\}$ 1.2398×10^{-6}	$\{h/m_u c\}$ 1.3310×10^{-15}	$\{1/\{2R_\infty\}\}$ 4.5563×10^{-8}
1 Hz =	$\{h\}$ 6.6261×10^{-34}	$\{h/c^2\}$ 7.3725×10^{-51}	$\{1/c\}$ 3.3356×10^{-9}	1	$\{h/k\}$ 4.7992×10^{-11}	$\{h/e\}$ 4.1357×10^{-15}	$\{h/m_u c^2\}$ 4.4398×10^{-24}	$\{1/\{2R_\infty c\}\}$ 1.5198×10^{-16}
1 K =	$\{k\}$ 1.3807×10^{-23}	$\{k/c^2\}$ 1.5362×10^{-40}	$\{k/hc\}$ 69.504	$\{k/h\}$ 2.0837×10^{10}	1	$\{k/e\}$ 8.6174×10^{-5}	$\{k/m_u c^2\}$ 9.251×10^{-14}	$\{k/2R_\infty hc\}$ $3.1668(27)\times10^{-6}$
1 eV =	$\{e\}$ 1.6022×10^{-19}	$\{e/c^2\}$ 1.7827×10^{-36}	$\{e/hc\}$ 806554	$\{e/h\}$ 2.4180×10^{14}	$\{e/k\}$ 11604	1	$\{e/m_u c^2\}$ 1.0735×10^{-9}	$\{e/2R_\infty hc\}$ 0.036749
1 u =	$\{m_u c^2\}$ 1.4924×10^{-10}	$\{m_u\}$ 1.6605×10^{-27}	$\{m_u c/h\}$ 7.5130×10^{14}	$\{m_u c^2/h\}$ 2.2523×10^{23}	$\{m_u c^2/k\}$ 1.0809×10^{13}	$\{m_u c^2/e\}$ 931.49×10^{6}	1	$\{m_u c/2R_\infty h\}$ 3.4232×10^{7}
1 hartree =	$\{2R_\infty hc\}$ 4.3597×10^{-18}	$\{2R_\infty h/c\}$ 4.8509×10^{-35}	$\{2R_\infty\}$ 2.1947×10^{7}	$\{2R_\infty c\}$ 6.5797×10^{15}	$\{2R_\infty hc/k\}$ 3.1577×10^{5}	$\{2R_\infty hc/e\}$ 27.211	$\{2R_\infty h/m_u c\}$ 2.9213×10^{-8}	1

References

[1] *Explanatory Supplement to the Astronomical Almanac*, edited by P. K. Seidelmann (University Science Books, Mill Valley, CA, 1992).

[2] H. Moritz, Geodetic Reference System 1980 [*Bull. Geodesiq.* **62**, 348 (1988)].

[3] J. Laskar, Astron. Astrophys. **157**, 59 (1986).

[4] E. R. Cohen and B. N. Taylor, The 1986 adjustment of the fundamental physical constants [*Rev. of Mod. Phys.*, **59**, 1121 (1987)]; *CODATA Bulletin*; No. 63, November, 1986 (Pergamon, Oxford/New York).

[5] B. N. Taylor and E. R. Cohen, Recommended Values of the Fundamental Physical Constants: A Status Report [J. Res. Natl. Inst. Stand. Technol. **95**, 497 (1990)].

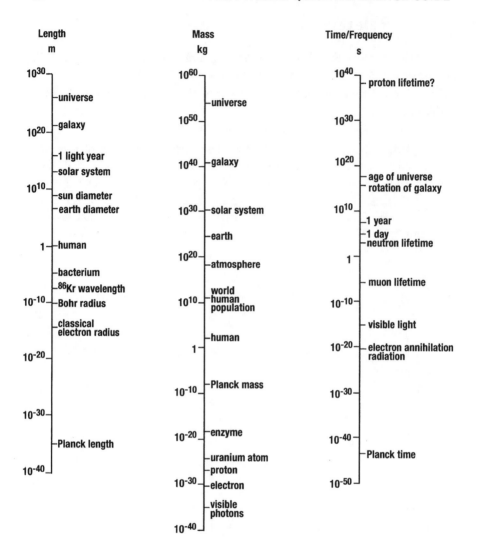

Length
m

10^{30}
—universe
10^{20}—galaxy
—1 light year
—solar system
10^{10}—
—sun diameter
—earth diameter
1—human
—bacterium
—^{86}Kr wavelength
10^{-10}—Bohr radius
—classical
electron radius
10^{-20}—
10^{-30}—
—Planck length
10^{-40}—

Mass
kg

10^{60}—
—universe
10^{50}—
—galaxy
10^{40}—galaxy
10^{30}—solar system
—earth
10^{20}—atmosphere
—world
10^{10}—human
population
—human
1—
—Planck mass
10^{-10}—
10^{-20}—enzyme
—uranium atom
—proton
10^{-30}—electron
—visible
photons
10^{-40}—

Time/Frequency
s

10^{40}—proton lifetime?
10^{30}—
10^{20}—
—age of universe
—rotation of galaxy
10^{10}—
—1 year
—1 day
—neutron lifetime
1—
—muon lifetime
10^{-10}—
—visible light
10^{-20}—electron annihilation
radiation
10^{-30}—
10^{-40}—
—Planck time
10^{-50}—

5

PHYSICS FORMULARY

5.1 Mechanics

5.1.1 Dynamics of particles and rigid bodies

Newton's laws of motion apply strictly to interacting point particles, but are extended to differential volume elements of finite bodies.

1. *Inertial Frame:* A Newtonian inertial frame is a coordinate system in which the velocity v of a particle is constant in the absence of forces, $F = 0$.

   "Every body continues in its state of rest, or of uniform motion in a right line, unless it is compelled to change that state by forces impressed on it."

   A body moving with velocity v (including a body at rest, $v = 0$) will not change its velocity unless acted upon by a force.

2. *Force and Momentum:* *"The change of motion [momentum] is proportional to the motive force impressed; it is made in the direction of the right line in which that force is impressed."*

   The rate of change of the momentum of a body, $p = mv$, is proportional to the force F acting on that body. The constant of proportionality is $k = 1$ in a coherent system of units.

   $$F = \frac{dp}{dt} = \frac{d(mv)}{dt}.$$

3. *Action and Reaction:* *"To every action there is always opposed an equal reaction: or, the mutual actions of two bodies upon each other are always equal, and directed to contrary parts."* The action of particle

1 on particle 2 is equal in magnitude and opposite in direction to the action of particle 2 on particle 1: $F_{12} = -F_{21}$.

Statics: Forces in balance, no acceleration: $\sum_i F_i = 0$.

d'Alembert's Principle states that a dynamic system can be reduced to an equivalent static system by replacing the acceleration a_i of each mass m_i by a force $F_i^a = -m_i a_i$. In a constrained system with impressed forces F_i, the effective forces, $F_i + F_i^a$, are in equilibrium with the forces of constraint: $F_i^c = F_i + F_i^a$.

Non-inertial Forces: In a coordinate frame Σ that is rotating at constant angular velocity ω with respect to an inertial frame Σ',

$$F = m\frac{d'^2 r'}{dt^2} = m\left[\frac{d^2 r}{dt} + 2\omega \times \frac{dr}{dt} + \omega \times (\omega \times r)\right]$$

$$m\frac{d^2 r}{dt^2} = F - 2m\omega \times v + m\omega^2 r_\perp,$$

where r' and $d'r'/dt$ are the position and velocity vectors in the inertial frame and r and $dr/dt = v$ are in the moving frame and $r_\perp$ is the component of r perpendicular to ω. The term $-2m\omega \times dr/dt$ is the *Coriolis force* and the term $-m\omega \times (\omega \times r) = m\omega^2 r_\perp$ is the *centrifugal force.*

Impulse: The impulse of a force,

$$I = \int_{t_1}^{t_2} F\, dt = m(v_2 - v_1),$$

measures the change in momentum.

Work: The work done is given by

$$W = \int_{s_1}^{s_2} F \cdot ds,$$

where s_1 and s_2 are two positions along the path s. The change in kinetic energy is

$$W = \tfrac{1}{2}mv_2^2 - \tfrac{1}{2}mv_1^2 = I \cdot \tfrac{1}{2}(v_1 + v_2).$$

Potential Energy: If the work done does not depend on the path but only on the initial and final positions, the force is said to be conservative. In this case $\nabla \times F = 0$ and $F = -\nabla \phi$ defines the potential energy ϕ. The total mechanical energy is

$$U = \tfrac{1}{2}mv^2 + \phi(x, y, z).$$

Rigid Body: The distance between any two points within a rigid body remains fixed. Six numbers suffice to locate a point in such an object:

three to define the position of a fixed point (origin point) in the body and three more to define the orientation of a coordinate system fixed in the body. There are two types of motion: translation and rotation.

The total mass is

$$M = \int_V \rho(x, y, z)\, dV,$$

where ρ is the density and V is the volume. The center of mass is

$$\boldsymbol{R} = \frac{1}{M} \int_V \rho(x, y, z)\boldsymbol{r}\, dV$$

and the translational motion is described by

$$\boldsymbol{F} = M\ddot{\boldsymbol{R}},$$

where $\boldsymbol{F}$ is the resultant of the applied external forces:

$$\boldsymbol{F} = \sum_i \boldsymbol{F}_i.$$

For an instantaneous angular velocity $\boldsymbol{\omega}$, the velocity of any point in the rigid body with position $\boldsymbol{r}_p$ relative to an origin on the axis of rotation is

$$\boldsymbol{v}_p = \boldsymbol{\omega} \times \boldsymbol{r}_p.$$

The *angular acceleration* is $\boldsymbol{\alpha} = \dot{\boldsymbol{\omega}}$.

The total *moment of momentum* or *angular momentum* with respect to a fixed origin of coordinates (whether inside or outside the body) is given in Cartesian coordinates by

$$\begin{aligned}
\boldsymbol{L} = &\, \boldsymbol{i}(\omega_x[I_{yy} + I_{zz}] - \omega_y I_{yx} - \omega_z I_{zx}) \\
&+ \boldsymbol{j}(-\omega_x I_{xy} + \omega_y[I_{xx} + I_{zz}] - \omega_z I_{zy}) \\
&+ \boldsymbol{k}(-\omega_x I_{xz} - \omega_y I_{yz} + \omega_z[I_{xx} + I_{yy}]).
\end{aligned}$$

Moments of Inertia:

$$I_{xx} = \int_V (y^2 + z^2)\rho\, dV, \qquad I_{yy} = \int_V (x^2 + z^2)\rho\, dV,$$

$$I_{zz} = \int_V (x^2 + y^2)\rho\, dV.$$

Products of Inertia:

$$I_{xy} = I_{yx} = \int_V xy\rho\, dV, \qquad I_{yz} = I_{zy} = \int_V yz\rho\, dV,$$

$$I_{zx} = I_{xz} = \int_V zx\rho\, dV.$$

Table 5.1 Transformation properties under rotation, inversion, and time reversal. (S = scalar; V = vector; PV = pseudovector; T = tensor).

Physical quantity	Type	Rotation (tensorial rank)	Space inversion $(\boldsymbol{x}, t) \to (-\boldsymbol{x}, t)$	Time reversal $(\boldsymbol{x}, t) \to (\boldsymbol{x}, -t)$
Mechanical				
Coordinate, $\boldsymbol{x}$	V	1	Odd	Even
Velocity, $\boldsymbol{v}$	V	1	Odd	Odd
Momentum, $\boldsymbol{p}$	V	1	Odd	Odd
Angular momentum, $\boldsymbol{L} = \boldsymbol{x} \times \boldsymbol{p}$	PV	1	Even	Odd
Force, $\boldsymbol{F}$	V	1	Odd	Even
Torque, $\boldsymbol{x} \times \boldsymbol{F}$	PV	1	Even	Even
Energy, E, $p^2/2m$, $U(\boldsymbol{x})$	S	0	Even	Even
Electromagnetic				
Charge, charge density, q, ρ	S	0	Even	Even
Current, current density, $\boldsymbol{I}$, $\boldsymbol{j}$	V	1	Odd	Odd
Electric field, $\boldsymbol{E}$	V	1	Odd	Even
Polarization, $\boldsymbol{P}$	V	1	Odd	Even
Displacement, $\boldsymbol{D}$	V	1	Odd	Even
Magnetic induction, $\boldsymbol{B}$	PV	1	Even	Odd
Magnetization, $\boldsymbol{M}$	PV	1	Even	Odd
Magnetic field, $\boldsymbol{H}$	PV	1	Even	Odd
Poynting vector, $\boldsymbol{S} = \boldsymbol{E} \times \boldsymbol{H}$	V	1	Odd	Odd
Maxwell stress tensor, T, $T_{\alpha\beta}$	T	2	Even	Even

By a proper choice of the axes of the coordinate system the products of inertia can be reduced to $I_{xy} = I_{yx} = I_{yz} = I_{zy} = I_{zx} = I_{xz} = 0$. Then $I_{xx} = MR_x^2$, $I_{yy} = MR_y^2$, and $I_{zz} = MR_z^2$ are the principal moments of inertia, and R_x, R_y, and R_z are the *radii of gyration* about the x-, y-, and z-axes, respectively.

The *torque* is

$$T = \sum_i r_i \times F_i,$$

where r_i is the position vector, with respect to the origin, of the point of application of the force F_i.

Newton's second law for the motion of a rigid body about a fixed origin is

$$T = \dot{L}.$$

If the fixed origin is taken as the center of mass, the translational motion of the body is simply that of the center of mass.

In a fixed (inertial) reference frame, the elements of the moment of inertia tensor, I_{xx}, I_{xy}, etc., change with time, but with respect to a frame fixed in the body they are constants. Then, for motion about a fixed point, the axes rotate and

$$\dot{L} = i\dot{L}_x + j\dot{L}_y + k\dot{L}_z + \omega \times L,$$

where $\dot{L}_x$, $\dot{L}_y$, $\dot{L}_z$ are the derivatives taken with respect to the rotating coordinate system. Choosing principal axes so that the off-diagonal components vanish, $I_{xy} = I_{yz} = I_{zx} = 0$, yields *Euler's equations*

$$T = i[I_{xx}\dot{\omega}_x - (I_{yy} - I_{zz})\omega_y\omega_z] + j[I_{yy}\dot{\omega}_y - (I_{zz} - I_{xx})\omega_z\omega_x]$$
$$+ k[I_{zz}\dot{\omega}_z - (I_{xx} - I_{yy})\omega_x\omega_y].$$

5.1.2 Special relativity

Lorentz transformations: Transformation from an inertial reference frame Σ to an inertial frame Σ' moving with respect to Σ with velocity βc in the $+x$ direction:

$$\gamma = (1 - \beta^2)^{-1/2}, \quad \beta = \tanh\theta, \quad \gamma = \cosh\theta,$$

$$x' = \gamma(x - \beta ct) = x\cosh\theta - ct\sinh\theta, \quad y' = y, \quad z' = z,$$
$$ct' = \gamma(ct - \beta x) = -x\sinh\theta + ct\cosh\theta.$$

The inverse transformation ($\Sigma' \to \Sigma$) reverses the sign of β.

A body moving with velocity v in Σ moves with velocity v' in Σ':

$$\tanh\phi_x = v_x/c, \qquad \tanh\phi'_{x'} = v'_{x'}/c,$$

$$v'_{x'} = \frac{v_x - c\beta}{1 - \beta v_x/c} = c\tanh(\phi_x - \theta), \quad \phi'_{x'} = \phi_x - \theta,$$

$$v'_{y'} = \frac{v_y}{\gamma(1 - \beta v_x/c)} = \frac{v_y\cosh\phi_x}{\cosh\phi'_{x'}},$$

$$v'_{z'} = \frac{v_z}{\gamma(1 - \beta v_x/c)} = \frac{v_z\cosh\phi_x}{\cosh\phi'_{x'}}.$$

The transformation of forces on a moving particle:

$$F'_{x'} = F_x - \frac{\beta}{c - \beta v_x}(v_y F_y + v_z F_z),$$

$$F'_{y'} = \frac{c}{\gamma(c - \beta v_x)}F_y,$$

$$F'_{z'} = \frac{c}{\gamma(c - \beta v_x)}F_z.$$

The transformation of the electromagnetic field:

$$E'_{x'} = E_x, \qquad E'_{y'} = \gamma(E_y - \beta B_z), \qquad E'_{z'} = \gamma(E_z + \beta B_y),$$
$$B'_{x'} = B_x, \qquad B'_{y'} = \gamma(B_y + \beta E_z), \qquad B'_{z'} = \gamma(H_z - \beta E_y).$$

Energy: $E = \gamma mc^2 = mc^2 + \text{(kinetic energy)}$.

Energy-Momentum Transformation:

$$p'_x = \gamma(p_x - \beta E/c) = p_x \cosh\theta - E\sinh\theta/c,$$
$$p'_y = p_y, \qquad\qquad p'_z = p_z,$$
$$E' = \gamma(E - \beta p_x c) = -p_x c\sinh\theta + E\cosh\theta,$$
$$E'^2 - (\boldsymbol{p'}\cdot\boldsymbol{p'})c^2 = E^2 - (\boldsymbol{p}\cdot\boldsymbol{p})c^2 = m^2 c^4, \qquad\qquad p = \beta E/c.$$

5.1.3 General relativity

For the formal description of curved space, see Chapter 8.

Einstein Postulates:
1. Inertial mass is proportional to gravitational mass. (The proportionality constant is set equal to 1.)
2. The mass-energy density determines the metric tensor of space. Empty space is flat; the Lorentzian metric can be expressed as

$$ds^2 = -c^2 dt^2 + dx^2 + dy^2 + dz^2.$$

3. A particle moving in a gravitational field defined by the distribution of mass-energy moves on a four-space geodesic satisfying the differential equations

$$\frac{d^2 x^\alpha}{ds^2} + \Gamma^\alpha_{\mu\nu} \frac{dx^\mu}{ds} \frac{dx^\nu}{ds} = 0.$$

4. Light propagates along null geodesics, $ds = 0$.

Mass-Energy Density Tensor:

$$T_{\mu\nu} = (\rho c^2 + p)g_{\mu\sigma} g_{\nu\tau} \frac{dx^\sigma}{ds} \frac{dx^\tau}{ds} + pg_{\mu\nu},$$

where ρ is the matter density and p is the pressure (energy density per unit volume):

$$R_{\mu\nu} - \tfrac{1}{2}g_{\mu\nu}R + \Lambda g_{\mu\nu} = \frac{8\pi G}{c^4} T_{\mu\nu}.$$

Λ is the comsological constant introduced by Einstein to provide a static solution for a non-empty universe:

$$R_{\mu\nu} = \frac{8\pi G}{c^4}(T_{\mu\nu} - \tfrac{1}{2}g_{\mu\nu}T) + \Lambda g_{\mu\nu}.$$

Schwarzschild Line Element: Spherically symmetric external solution with a mass $m = Mc^2/G$ at the origin $(M = mG/c^2)$.
In spherically symmetric coordinates:

$$ds^2 = -c^2(1 - 2M/r)dt^2 + (1 - 2M/r)^{-1}dr^2 + r^2(d\theta^2 + \sin^2\theta d\phi^2).$$

In isotropic coordinates:

$$ds^2 = -c^2 \left(\frac{1 - M/2R}{1 + M/2R}\right)^2 dt^2$$
$$+ (1 + M/2R)^4[dR^2 + R^2(d\theta^2 + \sin^2\theta d\phi^2)].$$

Clock Transformation: TAI (*Temps Atomique International*) is a coordinate time scale whose unit is the SI second as realized in a geocentric frame by a cesium atomic clock at rest at sea-level on the rotating geoid.
a.   Transfer by portable clock: the coordinate time accumulated in moving the clock from point A to point B is

$$\Delta t = \frac{2\omega}{c^2}A_{\text{E}} + \int_A^B \left[1 - \frac{U(\boldsymbol{r})}{c^2} + \frac{v^2}{c^2}\right] d\tau,$$

where $\boldsymbol{r}$ is the vector of the clock position in the earth frame with origin at the center of the geoid; v is the speed of the clock in that frame; ω is the angular speed of rotation of the earth $(2\omega/c^2 = 1.623{\times}10^{-21}\,\text{s/m}^2)$; A_{E} is the equatorial projection of the area swept out by $\boldsymbol{r}$ in the earth frame, counted positive when the projected motion of $\boldsymbol{r}$ is eastward; $d\tau$ is the proper time of the moving clock; and $U(\boldsymbol{r})$ is the gravitational potential relative to the geoid.
b.   Transfer by electromagnetic signal: the coordinate time interval between emission (at point A) to reception (at point B) is

$$\Delta t = \frac{2\omega}{c^2}A_{\text{E}} + \frac{1}{c}\int_A^B ds,$$

where ds is the increment of proper length along the path, and A_{E} is the equatorial projection of the triangle with vertices at A, B, and the center of the geoid.

5.1.4   Cosmology

Hubble Constant: $H \equiv 100h\,\text{km}\,\text{s}^{-1}\,\text{Mpc}^{-1}$, $0.35 < h_\text{o} < 0.85$,
where the subscript $_\text{o}$ indicates the value at the present time. $(1\,\text{Mpc} = 3.086{\times}10^{22}\,\text{m} = 3.262{\times}10^6\,\text{ly})$, Hubble time, $1/H = (9.7779{\times}10^9/h)\,\text{yr}$, Hubble distance, $c/H \approx (9.25{\times}10^{25}/h)\,\text{m} \approx (3000/h)\text{Mpc}$.

Redshift: If λ is the observed wavelength, $\lambda_{\rm o}$ the "proper" (emitted) wavelength, the redshift is $z = (\lambda - \lambda_{\rm o})/\lambda_{\rm o}$. The recession velocity is

$$v/c = \frac{(1+z)^2 - 1}{(1+z)^2 + 1}.$$

The present temperature of the redshifted big-bang background radiation is $T_{\rm BB} = 2.726(5)\,{\rm K}$.

Friedman-Robertson-Walker metric: In a homogeneous, isotropic universe the metric takes the form

$$(ds)^2 = -c^2(dt)^2 + R(t)^2 \left[\frac{(dr)^2}{1 - kr^2} + r^2[(d\theta)^2 + \sin^2\theta(d\phi)^2] \right],$$

where r is a dimensionless coordinate and $k = +1, 0, -1$ for a closed (spherical), flat, or open (hyperboloidal) space, respectively. $R(t)$ is the scale factor for distances measured in a comoving frame. The mean curvature of space is $k/R^2(t)$. For $k = +1$, the volume of the closed space is $2\pi^2 R(t)^3$:

$$\frac{\ddot{R}}{R} = \frac{1}{3}[\Lambda - 4\pi G(\rho + 3p/c^2)],$$

$$\left(\frac{\dot{R}}{R}\right)^2 \equiv H^2 = \frac{1}{3}(8\pi G\rho + \Lambda) - \frac{kc^2}{R^2},$$

where ρ is the density of matter and p is the pressure of radiation. A nonempty universe cannot be static ($\dot{R} = 0$) unless $\Lambda > 0$.

Critical Density: $\rho_{\rm c} = 3H^2/8\pi G$, the mass density required to close the uniform isotropic universe, $\rho_{\rm c} \approx 1.88 \times 10^{-26} h^2 \,{\rm kg/m^3}$:

$$\Omega = \rho/\rho_{\rm c}, \quad \Omega_{\rm o} = \rho_{\rm o}/\rho_{\rm c}.$$

From large-scale velocity measurements ($> 100\,{\rm kpc} = 3 \times 10^5\,{\rm ly}$), $0.1 < \Omega < 0.4$; approximately 95 percent of this density is apparently unobserved ("dark") matter.

5.1.5 Elasticity

Solid bodies resist the deformation caused by external forces by producing internal forces. The force per unit area on a surface element within the body is called the *stress*; any stress is expressible as the combination of a *normal* stress acting perpendicularly to the surface and two *shear* stresses acting tangentially to the surface element. In a crystalline substance the response is anisotropic and both stress and *strain* (which describes the resulting deformation) will be second-order tensors.

In a general anisotropic medium, let the displacement of a particle from its initial position in a solid body be

$$s = i\xi + j\eta + k\zeta = \sum_j e_j s_j,$$

where the displacement components $\xi = s_1$, $\eta = s_2$, $\zeta = s_3$ are functions of space and time. The derivatives $\partial s_j / \partial x_k$ ($x_1 = x$, $x_2 = y$, $x_3 = z$) are the components of a covariant tensor of the second order. These nine elements may be reexpressed in terms of three antisymmetric components and six symmetric components. The three antisymmetric components,

$$\alpha_1 = \tfrac{1}{2}(\partial s_3/\partial x_2 - \partial s_2/\partial x_3), \qquad \alpha_2 = \tfrac{1}{2}(\partial s_1/\partial x_3 - \partial s_3/\partial x_1),$$
$$\alpha_3 = \tfrac{1}{2}(\partial s_2/\partial x_1 - \partial s_1/\partial x_2)$$

represent a *rotation*, while the six symmetric components represent a distortion or *strain*:

$$\mathbf{D} = \begin{pmatrix} \partial\xi/\partial x & \tfrac{1}{2}(\partial\xi/\partial y + \partial\eta/\partial x) & \tfrac{1}{2}(\partial\xi/\partial z + \partial\zeta/\partial x) \\ \tfrac{1}{2}(\partial\eta/\partial x + \partial\xi/\partial y) & \partial\eta/\partial y & \tfrac{1}{2}(\partial\eta/\partial z + \partial\zeta/\partial y) \\ \tfrac{1}{2}(\partial\zeta/\partial x + \partial\xi/\partial z) & \tfrac{1}{2}(\partial\zeta/\partial y + \partial\eta/\partial z) & \partial\zeta/\partial z \end{pmatrix}$$

$$= \begin{pmatrix} \epsilon_{11} & \tfrac{1}{2}\epsilon_{12} & \tfrac{1}{2}\epsilon_{13} \\ \tfrac{1}{2}\epsilon_{21} & \epsilon_{22} & \tfrac{1}{2}\epsilon_{23} \\ \tfrac{1}{2}\epsilon_{31} & \tfrac{1}{2}\epsilon_{23} & \epsilon_{33} \end{pmatrix}.$$

The stress is similarly expressed by a tensor:

$$\mathbf{S} = \begin{pmatrix} \tau_{11} & \tau_{12} & \tau_{13} \\ \tau_{21} & \tau_{22} & \tau_{23} \\ \tau_{31} & \tau_{32} & \tau_{33} \end{pmatrix},$$

where τ_{ii} are the normal stresses and $\tau_{jk} = \tau_{kj}$ are the torsional stresses.

For small deformations the strain is proportional to the stress (*Hooke's Law*, $\mathbf{S} = c\mathbf{D}$); hence,

$$\tau_{ij} = \sum_{k,l} c_{ijkl}\epsilon_{lk} \qquad \text{and} \qquad \epsilon_{lk} = \sum_{j,i} s_{lkji}\tau_{ij}.$$

c_{ijkl} are the *elastic constants* and s_{lkji} are the *elastic compliances*. The energy density of the deformation is

$$w = \tfrac{1}{2}\sum_{i,j} \tau_{ij}\epsilon_{ij} = \tfrac{1}{2}\sum_{ijkl} c_{ijkl}\epsilon_{lk}\epsilon_{ij}.$$

The 81 elastic constants are reduced to 21 independent components because of the symmetries expressed by

$$c_{ijkl} = c_{jikl} = c_{ijlk} = c_{lkji}.$$

The notation can be simplified by introducing a six-dimensional space with the correspondence $(i, j) \to I$: $(1, 1) \to 1$; $(2, 2) \to 2$; $(3, 3) \to 3$; $(2, 3), (3, 2) \to 4$; $(1, 3), (3, 1) \to 5$; $(1, 2), (2, 1) \to 6$. (The reduction from 9 terms to 6 terms is related to the absence of elastic forces associated with the 3 components of rigid body rotation.)

$$
\mathbf{D} = \begin{pmatrix} \epsilon_1 & \frac{1}{2}\epsilon_6 & \frac{1}{2}\epsilon_5 \\ \frac{1}{2}\epsilon_6 & \epsilon_2 & \frac{1}{2}\epsilon_4 \\ \frac{1}{2}\epsilon_5 & \frac{1}{2}\epsilon_4 & \epsilon_3 \end{pmatrix}
\qquad
\mathbf{S} = \begin{pmatrix} \tau_1 & \tau_6 & \tau_5 \\ \tau_6 & \tau_2 & \tau_4 \\ \tau_5 & \tau_4 & \tau_3 \end{pmatrix}.
$$

Then, with $I, K = 1, \ldots, 6$, Hooke's law becomes

$$
\tau_I = \sum_K C_{IK}\epsilon_K, \qquad \epsilon_J = \sum_L S_{JL}\tau_L,
$$

where $C_{IK} = C_{KI} = c_{ijkl}$ and $S_{JL} = S_{LJ} = s_{lkji}$.

The energy density in this notation is

$$
w = \tfrac{1}{2}\sum_{I,J} C_{IJ}\epsilon_I\epsilon_J.
$$

In crystalline material the 21 elastic constants are further restricted by crystal symmetry constraints. For example, in a cubic crystal, the equivalence of the axes requires $C_{11} = C_{22} = C_{33}$, $C_{44} = C_{55} = C_{66}$, and $C_{12} = C_{13} = C_{23}$ while the reflection symmetry requires $C_{4j} = C_{5j} = C_{6j} = 0$ for $j = 1, 2, 3$. If, in addition, $C_{12} = C_{11} - 2C_{44}$, the cubic crystal is elastically isotropic.

If all forces between molecules in a crystal were central, an additional symmetry constraint (Cauchy conditions) would apply:

$$
c_{ijkl} = c_{ikjl}, \qquad
\begin{array}{ll}
C_{44} = C_{23}, & C_{14} = C_{56}, \\
C_{55} = C_{13}, & C_{25} = C_{46}, \\
C_{66} = C_{12}, & C_{36} = C_{45},
\end{array}
$$

reducing the number of independent components to 15. To the extent that these conditions are not satisfied for actual materials, non-central forces must exist.

Isotropic Medium: In an isotropic medium the elastic constants have only two independent components:

$$
c_{ijkl} = \lambda\delta_{ij}\delta_{kl} + \mu(\delta_{ik}\delta_{jl} + \delta_{il}\delta_{jk}),
$$

where λ and μ are the *Lamé constants*. With $I, J = 1, 2, 3$ and $\alpha, \beta = 4, 5, 6$,

$$
C_{IJ} = \lambda + 2\mu\delta_{IJ}, \qquad C_{\alpha\beta} = \mu\delta_{\alpha\beta}, \qquad C_{I\alpha} = 0.
$$

Table 5.2 Elastic constants for some cubic crystals. The isotropy ratio is defined here by $R = (c_{1122} + 2c_{1212})/c_{1111} = (C_{12} + 2C_{44})/C_{11}$; it is equal to 1 for an elastically isotropic crystal.

Substance (300 K)	$C_{11}/(10^{10}\,\text{Pa})$	$C_{12}/(10^{10}\,\text{Pa})$	$C_{44}/(10^{10}\,\text{Pa})$	Isotropy ratio
Li (78 K)	1.48	1.25	1.08	2.3
Na	0.7	0.61	0.45	2.2
Cu	16.8	12.1	7.5	1.6
Ag	12.4	9.3	4.6	1.5
Au	18.6	15.7	4.2	1.3
Al	10.7	6.1	2.8	1.1
Pb	4.6	3.9	1.44	1.5
C (diamond)	107.6	12.5	57.6	1.2
Ge	12.9	4.8	6.7	1.4
Si	16.6	6.4	8	1.3
V	22.9	11.9	4.3	0.9
Ta	26.7	16.1	8.2	1.2
Nb	24.7	13.5	2.87	0.8
Fe	23.4	13.6	11.8	1.6
Ni	24.5	14.0	12.5	1.6
LiCl	4.94	2.28	2.46	1.5
NaCl	4.87	1.24	1.26	0.8
KF	6.56	1.46	1.25	0.6
RbCl	3.61	0.62	0.47	0.4
InSn	6.72	3.67	3.02	1.4
InAs	8.33	4.53	3.96	1.5
GaAs	11.88	5.38	5.94	1.5

Shear Modulus, μ (or G): $\tau_{ij} = \mu\epsilon_{ij} = \mu\left(\dfrac{\partial\xi_i}{\partial x_j} + \dfrac{\partial\xi_j}{\partial x_i}\right)$ $i \neq j$.

Young's Modulus, E (or Y): For a uniaxial stress, $\tau_{11} = \tau$, and all other stresses vanish, then $\tau = E\epsilon_{11}$ and $\epsilon_{22} = \epsilon_{33} = -\nu\epsilon_{11}$,

$$E = \frac{3\kappa\mu}{\lambda + \mu} = \frac{\mu(3\lambda + 2\mu)}{\lambda + \mu} = 2\mu(1 + \nu), \quad \nu = \frac{\lambda}{2(\lambda + \mu)},$$

where ν is the *Poisson ratio* (for $\mu > 0$ and $\lambda > 0$, $0 < \nu < \frac{1}{2}$).

5.2 Wave propagation

5.2.1 Plane waves

The one-dimensional wave equation

$$\frac{\partial^2}{\partial t^2} f(x,t) = v^2 \frac{\partial^2}{\partial x^2} f(x,t)$$

has the general solution

$$f(x,t) = g^+(x - vt) + g^-(x + vt),$$

where $g^+(z)$ and $g^-(z)$ are arbitrary functions representing wave forms traveling respectively in the positive and negative directions. With the initial conditions

$$f(x,0) = f(x), \qquad \frac{\partial f}{\partial t}\bigg|_{t=0} = f_t(x) = F'(x)$$

the solution is

$$f(x,t) = \tfrac{1}{2}\left[f(x+vt) + f(x-vt) + \frac{1}{v}[F(x+vt) - F(x-vt)]\right].$$

5.2.2 Wave propagation in fluids; linear acoustics

A wave of excess pressure p propagating in the positive x direction in a fluid of density ρ is subject to the equation

$$\frac{\partial^2 p}{\partial x^2} = \frac{1}{c^2}\frac{\partial^2 p}{\partial t^2}.$$

The speed of sound c is given by

$$c^2 = \frac{\partial p}{\partial \rho}\bigg|_S,$$

where S is the entropy. When viscosity η and thermal conductivity κ are included, a pure harmonic wave of frequency f has the form

$$p(x,t) = p_\circ e^{-\alpha x} \cos\left[\frac{2\pi f}{c}(x - ct) + \theta_\circ\right],$$

where $\theta_\circ$ is the phase at $x = 0$. The attenuation coefficient is

$$\alpha = \frac{\omega^2}{\rho c^3}\left[\frac{2\eta}{3} + \frac{\gamma - 1}{\gamma}\frac{\kappa}{2c_v}\right],$$

where $\omega = 2\pi f$, c_v is the heat capacity (per unit mass) at constant volume. Molecular relaxation absorption contributes an additional attenuation

$$\alpha(\omega) = \frac{2\alpha_m \omega^2 \tau^2}{1 + \omega^2 \tau^2},$$

where τ is the relaxation time and α_m is the absorption coefficient at $\omega = 1/\tau$, the frequency at which α/f has its maximum value.

5.2.3   Non-linear acoustics

For a plane wave propagating in the x-direction, the position $x(x_o, t)$ of a particle of fluid that was at x_o at time $t = 0$ (Lagrangian form of the wave equation) is given by

$$\frac{\partial^2 \xi}{\partial t^2} = \frac{c_o^2}{(1 + \partial \xi/\partial x_o)^{2+B/A}} \cdot \frac{\partial^2 \xi}{\partial x_o^2},$$

where $\xi = x - x_o$ and $c_o^2 = (\partial p/\partial \rho)|_S$,

$$A = \rho c_o^2, \qquad B = \rho_o^2 \frac{\partial^2 p}{\partial \rho^2}\Big|_S, \qquad B/A = \frac{\rho_o}{c_o^2} \frac{\partial c^2}{\partial \rho}\Big|_S$$

and the partial derivatives are taken at constant entropy S and evaluated at $\rho = \rho_o$, the equilibrium value of the fluid density. B/A is known as the parameter of non-linearity; for an ideal gas, $B/A = \gamma - 1$.

For harmonic waves with frequency $f = \omega/2\pi$, the solution of the Eulerian wave equation for the particle velocity is

$$u(x, t|f) = u_o \sin\left[\omega\left(t - \frac{x}{c_o}\left(1 + \frac{B}{2A}\frac{u}{c_o}\right)^{-(1+2A/B)}\right)\right].$$

5.2.4   Solitons

The Korteweg-deVries equation describes weakly nonlinear one-dimensional wave propagation:

$$u\frac{\partial u}{\partial x} + B\frac{\partial u}{\partial t} + C\frac{\partial^3 u}{\partial x^3} = 0 \qquad B, \ C \quad \text{constant.}$$

A solution of this equation is a moving pulse (soliton) whose fixed width and speed are amplitude dependent:

$$u(x, t) = A\operatorname{sech}^2 k(x - vt), \qquad v = A/3B, \quad k = \sqrt{A/12C}.$$

5.2.5 Wave propagation in elastic solids

In a infinite homogeneous elastic medium, simple harmonic waves are possible with

$$s = A \cos(k \cdot x - \omega t),$$

where A is a constant vector and k is the propagation constant of the wave, $k = \omega/v$, $k = k\gamma \!\!\!Æ$ is the propagation vector of the wave in the crystal with fixed direction cosines γ_m. Then, with $D_{ij} = \sum_{lm} \gamma_l c_{ilmj} \gamma_m$, one obtains the *eigen*condition

$$\sum_j \left(k^2 D_{ij} - \rho \omega^2 \delta_{ij} \right) A_j = 0.$$

Since D_{ij} is symmetric there are three real eigenvalues for k^2, each with an associated eigenvector A. Although the three eigenvectors are orthogonal, in general their orientation is not simply related to the direction of the propagation vector k.

For an infinite isotropic medium with Lamé constants λ and μ, the wave equation is

$$\rho \frac{\partial^2 r}{\partial t^2} = (\lambda + 2\mu) \nabla \nabla \cdot r - \nabla \times \nabla \times r.$$

There is a longitudinal or compressional wave with displacement in the direction of propagation and speed given by

$$\rho c_l^2 = \lambda + 2\mu,$$

and two transverse or shear waves with displacements at right angles to the direction of propagation and speed given by

$$\rho c_t^2 = \mu.$$

For a rod with transverse dimensions small compared to the wavelength, the speed of a compressional wave along the rod is

$$\rho c_y^2 = 2\mu + \frac{\lambda \mu}{\lambda + \mu} = \frac{\mu(3\lambda + 2\mu)}{\lambda + \mu} = E = \text{Young's modulus.}$$

5.3 Electricity and magnetism

5.3.1 Maxwell's equations

The equations are written in rationalized form corresponding to the convention of SI, which also defines current to be an independent base quantity. In SI the defining quantity for electromagnetic units is the permeability of vacuum, $\mu_\circ \equiv 4\pi \times 10^{-7} \, \text{N/A}^2$. The permittivity of vacuum is $\epsilon_\circ = 1/\mu_\circ c^2$. In contrast, the electrostatic system (*esu*) sets $4\pi \epsilon_\circ = 1$

while the electromagnetic system (*emu*) sets $\mu_o = 4\pi$ (thus canceling the appearance of 4π in the divergence equation). These systems, as well as the Gaussian (mixed) system, are not only "unrationalized" but are purely mechanical systems; all electrical quantities are derived quantities.

$$\nabla \cdot \boldsymbol{D} = \rho, \qquad \nabla \cdot \boldsymbol{B} = 0,$$

$$\nabla \times \boldsymbol{E} + \frac{\partial \boldsymbol{B}}{\partial t} = 0, \qquad \nabla \times \boldsymbol{H} = \boldsymbol{j} + \frac{\partial \boldsymbol{D}}{\partial t},$$

$$\boldsymbol{D} = \epsilon_o \boldsymbol{E} + \boldsymbol{P} = \epsilon \boldsymbol{E}, \qquad \boldsymbol{H} = \frac{1}{\mu_o}\boldsymbol{B} - \boldsymbol{M} = \frac{1}{\mu}\boldsymbol{B}.$$

Continuity: $\dfrac{\partial \rho}{\partial t} + \nabla \cdot \boldsymbol{j} = 0.$

Lorentz Force: $\boldsymbol{F} = q(\boldsymbol{E} + \boldsymbol{v} \times \boldsymbol{B}).$

Boundary Conditions between Two Media:

$$\boldsymbol{n} \cdot (\boldsymbol{D}_2 - \boldsymbol{D}_1) = \sigma, \qquad \boldsymbol{n} \times (\boldsymbol{E}_2 - \boldsymbol{E}_1) = 0,$$

$$\boldsymbol{n} \cdot (\boldsymbol{B}_2 - \boldsymbol{B}_1) = 0, \qquad \boldsymbol{n} \times (\boldsymbol{H}_2 - \boldsymbol{H}_1) = \boldsymbol{j}_{\text{surf}}.$$

5.3.2   Conservation of energy and momentum

For a charge q the rate of doing work is $W = q\boldsymbol{v} \cdot \boldsymbol{E}$, where $\boldsymbol{v}$ is the velocity of the charge. For a continuous distribution of charge and current in a volume τ,

$$W = \int_{\tau} \boldsymbol{j} \cdot \boldsymbol{E} \, d\tau.$$

The total energy density is given by

$$u = \tfrac{1}{2}(\boldsymbol{E} \cdot \boldsymbol{D} + \boldsymbol{B} \cdot \boldsymbol{H}).$$

The Poynting vector $\boldsymbol{S}$, represents energy flow and is given by

$$\boldsymbol{S} = \boldsymbol{E} \times \boldsymbol{H}.$$

It has the dimension of power per unit area. The electromagnetic momentum density is $\boldsymbol{g} = \boldsymbol{S}/c^2$.

5.3.3   Electrostatics

All expressions given here are in rationalized units; to express the relations in Gaussian electrostatic units set $4\pi\epsilon_o = 1$.

Table 5.3 Transformation between Gaussian and SI quantities.
To convert an equation from one system to the other replace the symbols
below by their counterparts consistently throughout the equation. Quantities that transform similarly, differing dimensionally only by powers of
length and/or time (e.g., charge, current, charge density), are grouped together. When, as is usually the case, the Gaussian equations are applied
in the cgs system, it is necessary to convert not only the equations but also
the units.

Quantity	Gaussian	SI
Speed of light	c	$1/\sqrt{\mu_0 \epsilon_0}$
Electric field, potential, voltage	$(\boldsymbol{E},\ \Phi,\ V)$	$(4\pi\epsilon_0)^{1/2}(\boldsymbol{E},\ \Phi,\ V)$
Displacement	$\boldsymbol{D}$	$(4\pi/\epsilon_0)^{1/2}\boldsymbol{D}$
Charge, charge density, current,	$(q,\ \rho,\ I,$	$(4\pi\epsilon_0)^{-1/2}(q,\ \rho,\ I,$
current density, polarization	$\boldsymbol{J},\ \boldsymbol{P})$	$\boldsymbol{J},\ \boldsymbol{P})$
Magnetic induction	$\boldsymbol{B}$	$(4\pi/\mu_0)^{1/2}\boldsymbol{B}$
Magnetic field	$\boldsymbol{H}$	$(4\pi\mu_0)^{1/2}\boldsymbol{H}$
Magnetization	$\boldsymbol{M}$	$(\mu_0/4\pi)^{1/2}\boldsymbol{M}$
Conductivity	σ	$\sigma\big/4\pi\epsilon_0$
Permeability	μ	μ/μ_0
Resistance, impedance	$(R,\ Z)$	$4\pi\epsilon_0(R,\ Z)$
Inductance	L	$4\pi\epsilon_0 L$ [a]
Capacitance	C	$(1/4\pi\epsilon_0)C$

[a] The Gaussian L is expressed here in an electrostatic unit (esu); for the Gaussian
    L expressed in an electromagnetic unit (emu), the conversion factor is $4\pi/\mu_0$.

a. Electric Potential:

$$V = \frac{1}{4\pi\epsilon_0}\sum_i \frac{q_i}{r_i}.$$

For a *charge distribution* with volume density ρ and surface density σ,

$$V = \frac{1}{4\pi\epsilon_0}\left(\int_\tau \frac{\rho}{r}\,d\tau + \int_S \frac{\sigma}{r}\,dS\right),$$

where r is the distance from the elements of volume $d\tau$ and of surface dS
to the field point. For an isotropic homogeneous medium with dielectric
constant $\epsilon = \epsilon_0\epsilon_r$, replace ϵ_0 by ϵ

$$\boldsymbol{E} = -\nabla V.$$

b. Poisson's Equation: $\nabla^2 V = -\rho/\epsilon_r.$

c. *Coulomb's Law for Two Point Charges:*

$$F_{12} = \frac{q_1 q_2}{4\pi\epsilon r_{12}^3} r_{12}.$$

d. *Electric Field Due to an Array of Point Charges:*

$$E = \frac{1}{4\pi\epsilon} \sum_i \frac{q_i}{r_i^3} r_i,$$

where r_i is the radius vector from the point charge to the field point.

A *dipole* is formed from two charges $-q$ and $+q$ separated by a distance l; the electric dipole moment is $p = ql$. The vector l is directed from the negative to the positive charge. The description of the electromagnetic field in terms of charges, dipoles, quadrupoles, etc., refers to mathematical (point) dipoles (p finite, $l \to 0$). Let r be the radius vector from the dipole to the field point ($r \gg l$):

$$V = \frac{p \cdot r}{4\pi\epsilon r^3},$$

$$E = \frac{1}{4\pi\epsilon} \left(\frac{3(p \cdot r)r}{r^5} - \frac{p}{r^3} \right),$$

$$E_r = \frac{2p\cos\theta}{4\pi\epsilon_o r^3}, \qquad E_\theta = \frac{p\sin\theta}{4\pi\epsilon_o r^3},$$

$$E = \frac{p}{4\pi\epsilon_o r^3}(1 + 3\cos^2\theta)^{1/2}.$$

In a *dielectric medium* $D = \epsilon_o E + P$ where $P = \lim_{\tau \to 0} \left(\frac{1}{\tau} \sum_i p_i \right)$ is the polarization, τ is a volume element (small with respect to macroscopic dimensions, but large on an atomic scale), and p_i is the dipole moment of the ith molecule (or atom) in the volume element.

e. *Gauss's Theorem:* The displacement flux

$$\Phi = \int_S D \cdot n \, dS = Q,$$

where Q is the total charge enclosed within the surface S, and n is the unit vector normal to the surface element dS.

f. *Electrostatic Energy:* The energy of the electrostatic interaction of an array of point charges is

$$U = \frac{1}{2} \sum_i q_i V_i = \frac{1}{8\pi\epsilon} \sum_{i \neq j} \frac{q_i q_j}{|r_i - r_j|},$$

where V_i is the potential at the position of the charge q_i due to all other charges.

For a capacitor, $W = \frac{1}{2}CV^2$ where C is the capacitance. In terms of surface and volume density of charge,

$$W = \frac{1}{2}\int_\tau \rho V \, d\tau + \frac{1}{2}\int_S \sigma V \, dS = \frac{1}{2}\int_\tau \boldsymbol{D}\cdot\boldsymbol{E} \, d\tau.$$

Circular cylinder of radius a and infinite length charged uniformly at linear density ζ (surface density $\zeta/2\pi a$):

$$V = \begin{cases} V_\circ, & r \le a \\ V_\circ + \dfrac{\zeta}{2\pi\epsilon}\ln\dfrac{a}{r}, & r \ge a \end{cases},$$

$$\boldsymbol{E} = \begin{cases} 0, & r \le a \\ \dfrac{\zeta}{2\pi\epsilon r^2}\boldsymbol{r}, & r \ge a \end{cases},$$

where $\boldsymbol{r}$ is the radius vector to the field point.

Infinite plane charged uniformly to surface density σ:

$$V_1 - V_2 = \frac{\sigma}{2\epsilon}\left(|x_2| - |x_1|\right),$$

$$\boldsymbol{E} = \frac{\sigma}{2\epsilon}\boldsymbol{n},$$

where x_1 and x_2 are the perpendicular distances from the plane to points 1 and 2.

The field between *two infinite parallel plates*, uniformly and oppositely charged, separated by a distance d:

$$\boldsymbol{E} = \frac{\sigma}{\epsilon}\boldsymbol{n},$$

where the unit vector $\boldsymbol{n}$ is perpendicular to the planes, and points from the positive to the negative charges. The potential difference between the two planes is $V_2 - V_1 = \sigma d/\epsilon$.

Sphere of radius R, whose charge Q is uniformly distributed over the surface ($\sigma = Q/4\pi R^2$):

$$r \le R, \quad V(r) = \frac{Q}{4\pi\epsilon_\circ R}, \quad \boldsymbol{E} = 0,$$

$$r \ge R, \quad V(r) = \frac{Q}{4\pi\epsilon_\circ r}, \quad \boldsymbol{E} = \frac{Q}{4\pi\epsilon_\circ r^3}\boldsymbol{r}.$$

If a sphere of radius R is charged uniformly throughout its volume with a volume charge density $\rho = Q/(4\pi R^3/3)$:

$$
V(\boldsymbol{r}) = \begin{cases} \dfrac{\rho}{2\epsilon_\circ}[R^2 - \dfrac{1}{3}r^2], & r \leq R \\ \dfrac{Q}{4\pi\epsilon_\circ r}, & r \geq R \end{cases} ,
$$

$$
\boldsymbol{E}(\boldsymbol{r}) = \begin{cases} \dfrac{\rho}{3\epsilon_\circ}\boldsymbol{r}, & r \leq R \\ \dfrac{Q}{4\pi\epsilon_\circ r^3}\boldsymbol{r}, & r \geq R \end{cases} .
$$

In these last two cases the potential and the field for $r > R$ is the same as for a point charge at the center of the sphere.

The corresponding gravitational potentials and fields are obtained by the substitutions $|Q| \to M$, $1/4\pi\epsilon_\circ \to -G$.

5.3.4 Capacitance

In these expressions $\epsilon = \epsilon_\circ\epsilon_r$ is the dielectric constant of the medium; for air $\epsilon_r \approx 1$. Edge effects are neglected. A is the area, d is the separation, R, R_1, R_2 are radii. For wires, a is the radius, l is the length, h is the height above ground.

$$C$$

a. *Parallel plates* : $\epsilon A/d$,

b. *Isolated disk* : $8\epsilon R$,

c. *Isolated sphere* : $4\pi\epsilon R$,

d. *Concentric Spheres* : $4\pi\epsilon R_2 R_1/(R_2 - R_1)$,

e. *Coaxial Cylinders* : $2\pi\epsilon l/\ln(R_2/R_1)$,

f. *Single straight wire parallel to the ground:*

$$
2\pi\epsilon_\circ l\left[\ln\frac{2h}{a} + \ln\frac{l + \sqrt{l^2 + a^2}}{l + \sqrt{l^2 + 4h^2}}\right]^{-1},
$$

g. *Two horizontal parallel wires distance h above the ground:*

$$
\pi\epsilon_\circ l\left[\ln\frac{d}{a} - \frac{d^2}{8h^2}\right]^{-1} \quad (a, d \ll l).
$$

5.3.5 Electric currents, Kirchhoff's laws

To analyze the currents and potential differences in an electrical mesh consisting of p segments (branches) and m nodes (junction of 3 or more branches), a definite direction (e.g., counter-clockwise) for a circuit must be established as the positive direction. In any loop, a current in that direction is taken to be positive and an electromotive force $\mathcal{E}_k$ of the current sources is positive if it produces a current in that direction. There are $p - m + 1$ independent closed circuits in a mesh.

First Law: The algebraic sum of the currents entering a node is zero.

$$\sum_j I_j = 0.$$

Second Law: In every closed circuit in the mesh, with current I_j, resistance R_j, and electromotive force $\mathcal{E}_j$, in the jth branch,

$$\sum_j (\mathcal{E}_j - I_j R_j) = 0.$$

The first law implies $m - 1$ independent equations and the second law implies $p - m + 1$ additional independent equations in the determination of the current in each of the p branches.

The first law may be satisfied by defining currents $I^{(m)}$ circulating in a defined direction for each independent closed circuit. The current I_k in the kth branch is $I_k = \sum \pm I^{(m)}$ where the sum is over the circulating currents in those independent closed circuits that contain the kth branch.

5.3.6   Magnetostatics

a. *Ampère's Law:* Magnetic induction $\boldsymbol{B}$ is defined by

$$d\boldsymbol{F} = d\boldsymbol{I} \times \boldsymbol{B} = I d\boldsymbol{l} \times \boldsymbol{B},$$

where $d\boldsymbol{l}$ is a vector element of the conductor pointing in the direction of the current flow in that conductor.

b. *Biot-Savart Law:*

$$d\boldsymbol{B} = \frac{\mu_o I}{4\pi} \frac{d\boldsymbol{l} \times \boldsymbol{r}}{r^3},$$

$$\boldsymbol{B}(\boldsymbol{r}) = \frac{\mu_o}{4\pi} \int_\tau \frac{\boldsymbol{j}(\boldsymbol{r}') \times (\boldsymbol{r} - \boldsymbol{r}')}{|\boldsymbol{r} - \boldsymbol{r}'|^3} d\tau,$$

where $\boldsymbol{j}(\boldsymbol{r}')$ is the current density, and τ is the volume element at the point $\boldsymbol{r}'$.

c. *Vector Potential:* $\boldsymbol{B}(\boldsymbol{r}) = \nabla \times \boldsymbol{A}(\boldsymbol{r}).$
In the Coulomb gauge, $\nabla \cdot \boldsymbol{A} = 0$. Since for any scalar Ψ, $\nabla \times \nabla \Psi = 0$, the magnetic field is unchanged by the transformation $\boldsymbol{A} \to \boldsymbol{A} + \nabla \Psi$. If $\nabla \cdot \boldsymbol{A}' = g$, then $\boldsymbol{A} = \boldsymbol{A}' - \nabla \Psi$, where $\nabla^2 \Psi = g$, gives the transformation of the vector potential to the Coulomb gauge.

The vector potential satisfies the Poisson equation,

$$\nabla^2 A + \mu_0 j = 0,$$

with the solution

$$A(r) = \frac{\mu_0}{4\pi} \int \frac{j(r')}{|r - r'|} \, d\tau.$$

This is the basis for the calculation of the magnetic field generated by a given distribution of currents.

d. Magnetic Moment: If a current distribution is localized in a space small compared to the scale length in observations of it effects, it is useful to define the *magnetization*

$$M(r) = \frac{\mu_0}{4\pi} \cdot \frac{1}{2} \left(r \times j(r) \right)$$

and the *magnetic moment*

$$m = \frac{\mu_0}{4\pi} \int \frac{1}{2} r' \times j(r') \, d\tau.$$

If the current is confined to a plane closed circuit of a conductor of negligible cross section carrying a current I,

$$m = (\mu_0/4\pi)I \oint \frac{1}{2} r \times ds = (\mu_0/4\pi)ISn,$$

where S is the area of the circuit and n is the normal unit vector.

Potential Energy: $\quad U = -m \cdot B.$
This is not the total energy of the system; it does not include the energy required to maintain the current flow in the circuit.

Torque: $\quad m \times B$

e. Magnetization in Matter: $\quad M(r) = \sum_i N_i \langle m_i \rangle,$

where $\langle m_i \rangle$ is the average magnetic moment in a small volume at the point r; N_i is the average number of molecules or atoms of type i per unit volume.
The effective current density due to magnetization:

$$j_M = \frac{\mu_0}{4\pi} \nabla \times M.$$

The magnetic field is defined by $H = B/\mu_0 - M$.

f. Permeability, μ: $\quad B = \mu H = \mu_0 \mu_r H, \qquad \mu_r = 1 + M/\mu_0 H.$

For diamagnetic substances $\mu_r - 1 \approx -10^{-5}$; for paramagnetic substances $\mu_r - 1 \approx 10^{-5} - 10^{-4}$. For ferromagnetic substances μ_r is large and depends on $\boldsymbol{H}$ and on the history of magnetization; μ is then defined by $1/\mu = 1/\mu_\circ - \partial M/\partial B$. (If the magnetization is not parallel to the induction, μ is a second rank tensor rather than a scalar.) The $\boldsymbol{B}$ versus $\boldsymbol{H}$ curve exhibits hysteresis.

Table 5.4 Saturation magnetization of high-permeability materials. These values are only representative; actual values vary with composition, treatment, and mechanical working of the material.

Material		$M/(\text{MA/m})$		Footnote
		20 °C	0 K	
Iron	Fe	1.714	1.732	a
Cobalt	Co	1.422	1.445	a
Nickel	Ni	0.484	0.5088	a
Metglas	80Fe, 20B	1.27		b
	80Fe, 16P, 3C, B	1.36		b
78-Permalloy	78Ni, 22Fe	0.851		c
Supermalloy	79Ni, 16Fe, 5Mo	0.692		c
Permendur	50Fe, 50Co	1.91		b
	65Fe, 35Co		1.958	a
	Sm Co_5	0.780		b
Ferrite	$\text{Mn Fe}_2\text{O}_4$	0.358		a
Magnetite	Fe_3O_4	0.485		a
Nickel ferrite	Ni Fe O_2	0.238		a
Cobalt ferrite	Co Fe O_2	0.42		d
YIG	$\text{Y Fe}_5\text{O}_{12}$	0.139		c
Gd iron garnet	$\text{Gd Fe}_5\text{O}_{12}$	0.014		c

[a] R. M. Bozorth, *Ferromagnetism* (van Nostrand, New York, 1951).

[b] D. Jiles, *Introduction to Magnetism and Magnetic Materials* (Chapman and Hall, New York, 1991).

[c] C-W. Chen, *Magnetism and Metallurgy of Soft Magnetic Materials* (Elsevier, New York, 1977).

[d] B. D. Cullity, *Introduction to Magnetic Materials* (Addison-Wesley, Reading, MA, 1971).

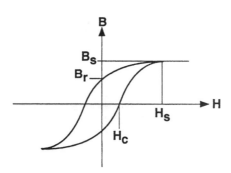

B_{sat} and H_{sat} are the *saturation* values. B_{r} is the remnant magnetic induction (retentivity); H_{c} is the coercive field (coercivity).

5.3.7   Magnetic fields due to currents

The "Right-Hand" Rules for Magnetic Fields and Currents:

1. The magnetic field generated by a current flowing in a wire circles around the wire in the direction of the fingers of the right hand when the extended thumb points in the direction of the positive current flow.

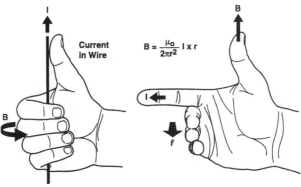

Table 5.5 Properties of high-permeability materials. The values given here are representative; actual values can be extremely sensitive to composition, heat treatment, and mechanical working of the material. From Ref. 9.

Material, [composition (%)]	Relative permeability μ_r(Initial)	μ_r(Final)	Coercivity $H_c/\mathrm{A\,m^{-1}}$	Retentivity B_r/T	Saturation B_{max}/T	Resistivity $\rho/\mu\Omega\,\mathrm{m}$
Iron						
Pure (annealed)	25 000	350 000	4	1.2	2.15	0.097
Swedish	250	5 500	80	1.3	2.1	0.1
Cast	100	600	360	0.53	2.0	0.3
hot rolled, (96Fe,4Si)	500	7 000	24	0.7	2.0	0.5
Rhometal (64Fe,36Ni)	1000	5 000	40	0.36	1.0	0.9
Permalloy45 (55Fe,45Ni)	2500	25 000	24		1.6	0.45
Mumetal[a]	20 000	100 000	4	0.6	0.72	0.25–0.5
Supermalloy (79Ni,5Mo,16Fe)	100 000	1 000 000	0.16		0.8	0.6
HyMu80 (80Ni,20Fe)	20 000	100 000	4		0.87	0.57
Alfenol (84Fe,16Al)	3 450	116 000	2	0.38	0.7825	1.5
Permendur (48-49Fe,50Co,1-2V)	800	4 500	160	1.4	2.4	0.26
Sendust (cast) (85Fe,10Si,5Al)	30 000	120 000	4	0.5	1.0	0.6–0.8
Ferroxcube-III (Mn-Zn-Ferrite)	1 000	1 500	8	0.1	0.3	$> 10^4$
Ferroxcube101 (Ni-Zn-Ferrite)	1 100		14	0.11	0.23	$> 10^5$

[a] Composition is variable (71-78Ni,4-6Cu,0-2Cr,14-25Fe).

Table 5.6 Curie temperature.   The Curie temperature is sensitive to trace impurity content and to crystal structure (vacancies, interstitials, dislocations, and crystal size).

Material		Curie temperature		
		$T_C\big/{}^\circ\mathrm{C}$	$T_C\big/\mathrm{K}$	Footnote
Fe	(iron)	770	1043	a
Ni	(nickel)	358	631	a
Co	(cobalt)	1112 −1145	1385 −1418	a
		1145	1418	a
$Nd_2Fe_{14}B$		312	585	b
$Sm\,Co_5$		720	993	b
$Mn\,Fe_2O_4$	(manganin ferrite)	510	783	b
Fe_3O_4	(magnetite)	575	848	c
$Ni\,Fe_2O_4$	(nickel ferrite)	590	863	c
$Co\,Fe_2O_4$	(cobalt ferrite)	520	793	c
$Y\,Fe_5O_{12}$	(YIG)	280	553	c
$Gd\,Fe_5O_{12}$	(gadolinium iron garnet)	291	564	c

[a] R. M. Bozorth, *Ferromagnetism* (van Nostrand, New York, 1951).
[b] D. Jiles, *Introduction to Magnetism and Magnetic Materials* (Chapman and Hall, New York, 1991).
[c] C-W. Chen, *Magnetism and Metallurgy of Soft Magnetic Materials* (Elsevier, New York, 1977).

2. If the extended index finger of the right hand is in the direction of the current and the bent fingers point from the current element to the field point, the thumb will point in the direction of the induced magnetic field.

a. *Straight Wire Segment*:

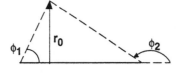

$$\boldsymbol{B} = \frac{\mu_0}{4\pi r_0^2}(\cos\phi_1 - \cos\phi_2)\boldsymbol{I}\times\boldsymbol{r}_0,$$

b. *Straight Wire, Infinite Length*: $B = \mu_0 I/2\pi r_0.$

c. Circular Loop: Field on the axis a distance h above the plane of the current loop.

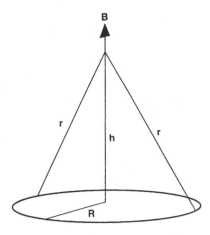

$$B = \frac{\mu_\circ I R^2}{2(R^2 + h^2)^{3/2}}.$$

The field is perpendicular to the plane of the loop.

d. Toroid: The field inside the toroid with relative permeability μ_r and N turns of wire carrying current I:

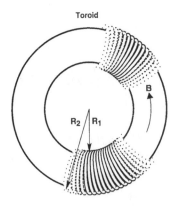

$$B = \frac{\mu_\circ \mu_r N I}{2\pi r},$$
$$H = N I / 2\pi r,$$
$$R_1 \leq r \leq R_2.$$

e. Solenoid with n Turns Per Unit Length: magnetic induction on the axis:

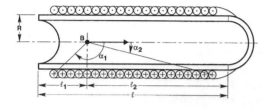

Solenoid

$$B = \tfrac{1}{2}\mu_\circ nI(\cos\alpha_2 - \cos\alpha_1),$$
for an infinitely long
solenoid: $B = \mu_\circ nI.$

5.3.8   Inductance

The expressions given here are approximate. Edge effects are neglected; the (low-frequency) field inside a wire is assumed to be uniform over the cross section; the pitch angle of the coil winding is assumed to be small. a is the radius of the wire. For coils R, R_1, R_2 are radii. l is a length, d is a separation. f = frequency, ρ = resistivity. L is the self-inductance, M is the mutual inductance

$$\mu_\circ \equiv 0.4\pi\,\mu\mathrm{H/m} = 0.4\pi\,\mu\Omega{\cdot}\mathrm{s/m},$$

$$L = \frac{1}{\mu_\circ I^2}\int_\tau B^2\,d\tau; \qquad M_{12} = \mu_\circ \oint\oint \frac{d\boldsymbol{s}_1 \cdot d\boldsymbol{s}_2}{|\boldsymbol{r}_1 - \boldsymbol{r}_2|}.$$

In the following expressions, $\delta(x)$ gives the frequency dependence (skin effect) in wires; μ_r is the relative permeability of the wire. d_s is the skin depth ($d_\mathrm{s}^2 = 2\rho/\mu\omega$); $x = \sqrt{2}\,a/d_\mathrm{s} = a\sqrt{2\pi\mu f/\rho}$:

$$\delta(x) \approx [16e^{-x^2/8} + 2x^2] = \frac{1}{4}[e^{-a^2/4d_\mathrm{s}^2} + a^2/4d_\mathrm{s}^2]^{-1/2}$$

x	$\delta(x)$	x	$\delta(x)$	x	$\delta(x)$	x	$\delta(x)$
0.0	0.250	2.0	0.240	4.0	0.1715	8.0	0.088
0.5	0.250	2.5	0.228	5.0	0.139	10.0	0.070
1.0	0.249	3.0	0.211	6.0	0.116	12.0	0.059
1.5	0.247	3.5	0.191	7.0	0.100	14.0	0.050

for $x \geq 10$, $\delta \approx d_\mathrm{s}/2a.$

a. *Straight Wire, Circular Cross Section:*

$$L = \frac{\mu_o}{2\pi} l \left[\ln \frac{2l}{a} - 2 + \mu_r \delta(x) \right].$$

b. *Two Parallel Wires:* (opposing currents)

$$L = \frac{\mu_o}{\pi} l \left[\ln \frac{d}{a} - \frac{d}{l} + \mu_r \delta(x) \right].$$

This expression neglects the inductance of the wires that complete the circuit for the two main wires; if these wires are long their inductance may be found from (a) and added to this result. If the complete circuit is in the form of a rectangle, use the following expression, (c).

c. *Rectangular Circuit:* The sides of the rectangle are l_1 and l_2, the diagonal is $g = \sqrt{l_1^2 + l_2^2}$:

$$L = \frac{\mu_o}{\pi} \left[(l_1 + l_2) \left[\ln \frac{l_1 l_2}{a^2} - 2 + \mu_r \delta(x) \right] + 2(g + a) \right.$$
$$\left. - l_1 \ln \frac{g + l_1}{2a} - l_2 \ln \frac{g + l_2}{2a} \right]$$

and for a square of side l

$$L = \frac{2\mu_o l}{\pi} \left[\ln \frac{l}{a} + \frac{a}{l} + \mu_r \delta(x) - 0.774 \right].$$

d. *Circular Ring of Circular Cross Section (torus):* , a is the radius of the cross section, R is the mean radius of the ring: (approximate expression for $R \geq 5a$):

$$L = \mu_o R \left[\ln \frac{8R}{a} - 2 + \mu_r \delta(x) \right].$$

For a toroidal coil of N wires:

$$L = \mu_o R N^2 \left[\ln \frac{8R}{a} - 2 + \mu_r \delta(x) \right].$$

e. *Single Layer Solenoid:* (N is the number of turns, R is the radius of the coil measured to the center of the wires, and l is the length of the coil:

$$L = \pi \mu_o \frac{R^2 N^2}{l} \cdot K(R/l),$$

where $K(R/l)$ is given in the following table.

R/l	$K(R/l)$	R/l	$K(R/l)$	R/l	$K(R/l)$	R/l	$K(R/l)$
0.0	1.0000	0.5	0.6884	1.2	0.4816	3.0	0.2854
0.1	0.9201	0.6	0.6475	1.5	0.4292	4.0	0.2366
0.2	0.8499	0.8	0.5795	2.0	0.3654	5.0	0.2033
0.4	0.7351	1.0	0.5255	2.5	0.3198	10.0	0.1236

f. *Torus with Single-Layer Winding:* (a is the radius of the wires; R is the distance from the axis to the center of the wires):

$$L = \pi N^2 [R - \sqrt{R^2 - a^2}].$$

g. *Grounded Horizontal Wire:* (Earth forms the return circuit; h is the distance above ground.)

$$L = \frac{\mu_o}{2\pi} l \left[\ln \frac{2h}{a} - \frac{2h - a}{l} + \ln \frac{l + \sqrt{1 + \frac{a^2}{l^2}}}{l + \sqrt{l^2 + 4h^2}} - \sqrt{1 + \frac{a^2}{l^2}} + \mu_r \delta(x) \right].$$

h. *Mutual Induction of Two Grounded Parallel Wires:*

$$M = \frac{\mu_o}{2\pi} l \left[\ln \frac{l + \sqrt{l^2 + d^2}}{d} - \frac{\sqrt{l^2 + d^2} - d}{l} \right.$$

$$\left. - \ln \frac{l + \sqrt{l^2 + 4h^2 + d^2}}{\sqrt{4h^2 + d^2}} + \frac{\sqrt{l^2 + 4h^2 + d^2} - \sqrt{4h^2 + d^2}}{l} \right].$$

5.3.9   Impedance in ac circuits

Ohm's law: $V = V_o e^{i\omega t} = Z(\omega) I_o e^{i\omega t}.$
Impedance of vacuum: $Z_o = \sqrt{\mu_o/\epsilon_o} = \mu_o c = 376.73\,\Omega.$
Impedance of an inductance L: $Z = i\omega L.$
Impedance of a capacitance C: $Z = 1/i\omega C.$
Impedance per unit length of a flat conductor of width w at high frequencies (conductor thickness much greater than the skin depth d_s):

$$\hat{Z} = \frac{(1 + i)\rho}{w d_s} = (1 + i)\sqrt{\pi \mu \rho f}/w,$$

where $d_s = \sqrt{\rho/\pi \mu f}$, ρ is the resistivity, $\mu = \mu_o \mu_r$ is the permeability of the conductor, and $f = \omega/2\pi$ is the frequency.

For a resistance R, capacitance C, and inductance L in series with a voltage source $V(t)$, the current in the circuit is $I = dq/dt$, where q satisfies

$$L\frac{d^2 q}{dt^2} + R\frac{dq}{dt} + \frac{q}{C} = V.$$

For $V(t) = V_\circ e^{i\omega t}$, solutions of this equation are $q(t) = q_s e^{i\omega t} + q_t(t)$ and $I(t) = I_s e^{i\omega t} + I_t(t)$. The steady state solution is

$$I_s = i\omega q_s = V_\circ/Z,$$
$$Z = R + i(\omega L - 1/\omega C),$$
$$|Z| = \sqrt{R^2 + (\omega L - 1/\omega C)^2}.$$

With the initial conditions $q(0) = \bar{q}_\circ + q_s(0)$, $I(0) = I_\circ$, the transient solution can be one of three types, depending on the value of $\Delta^2 = R^2 - 4L/C$:

(a) Overdamped, $\Delta^2 > 0$, Δ real:

$$q_t(t) = \frac{1}{p_1 - p_2}[(I_\circ + p_1 \bar{q}_\circ)e^{-p_1 t} - (I_\circ + p_2 \bar{q}_\circ)e^{-p_2 t}]$$

with $p_1 = (R+\Delta)/2L$, $p_2 = (R-\Delta)/2L$. The circuit is unstable (p_1 or $p_2 < 0$) if L, R, C are not all of the same sign.

(b) Critically damped, $\Delta^2 = 0$, $p_1 = p_2 = p = R/2L = 2/RC$:

$$q_t(t) = [\bar{q}_\circ + (I_\circ + p\bar{q}_\circ)t]e^{-pt}.$$

(c) Underdamped, $\Delta^2 < 0$, $\omega_1 = \sqrt{-\Delta^2}/2L$:

$$q_t(t) = \left[\bar{q}_\circ \cos \omega_1 t + (I_\circ + p\bar{q}_\circ)\frac{\sin \omega_1 t}{\omega_1}\right]e^{-pt}.$$

The resonant frequency, at which $|Z(\omega)|$ is a minimum ($Z = R$), is $\omega_\circ = 1/\sqrt{LC}$; $\omega_1 = \omega_\circ \sqrt{1 - 1/4Q^2}$, where the "quality" of the circuit, Q, is given by

$$Q = \sqrt{L/C}/R = \omega_\circ L/R = 1/\omega_\circ RC.$$

5.3.10 Transmission lines (no loss)

For flat plates of width w separated by $d \ll w$ per unit length,

$$\hat{C} = \epsilon w/d, \qquad \hat{L} = \mu d/w, \qquad\qquad \epsilon = \epsilon_\circ \epsilon_r, \qquad \mu = \mu_\circ \mu_r,$$

where $\hat{C}$ and $\hat{L}$ are the capacitance and inductance per unit length. For a coaxial cable of inside radius r_1 and outside radius r_2:

$$\hat{C} = \frac{2\pi\epsilon}{\ln(r_2/r_1)}, \qquad \hat{L} = \frac{\mu}{2\pi} \ln(r_2/r_1),$$

$$\text{velocity (speed)} = v = \frac{1}{\sqrt{\hat{L}\hat{C}}} = \frac{1}{\sqrt{\epsilon\mu}} = \frac{c}{\sqrt{\epsilon_r\mu_r}},$$

$$\text{impedance per unit length} = \sqrt{\hat{L}/\hat{C}}$$

($\epsilon_r = 2 - 6$ for plastics and $4 - 8$ for typical porcelain and glasses).

5.3.11 Synchrotron radiation

For particles of charge e moving with speed $v = \beta c$ in a circular orbit of radius R:

$$\text{energy loss per revolution} = \frac{e^2}{3\epsilon_0 R}\gamma(\gamma^2 - 1)^{3/2},$$

where $\gamma = (1 - \beta^2)^{-1/2} = E/mc^2$.

For relativistic electrons ($\gamma \gg 1$) the energy loss per revolution is

$$\frac{\Delta E}{\text{rev}} = \frac{(88.463\,\text{keV·m})}{R}[E/\,\text{GeV}]^4.$$

This energy is radiated primarily into a cone with a half angle of order $1/\gamma$ whose axis is in the forward direction with respect to the instantaneous motion of the electron in its orbit. The frequency spectrum $I(\omega)$ of synchrotron radiation is approximately

$$\frac{dI(\omega)}{d\omega} \approx 3.25\alpha\hbar(\omega R/c)^{1/3}, \qquad\qquad \omega \ll \omega_c,$$

$$\approx \alpha\hbar\sqrt{\frac{\pi\omega R}{\gamma c}}e^{-\omega/\omega_c}\left[1 + \frac{55\omega}{72\omega_c} + \cdots\right], \qquad \omega \gg \omega_c,$$

where $\alpha = e^2/4\pi\epsilon_0\hbar c = 1/137.036$ is the fine-structure constant and $\omega_c = 3\gamma^3 c/2R$. [But notations differ; Jackson (Ref. 3) defines $\omega_c = 3\gamma^3 c/R$.]

5.4 Hydrogen-like atom (non-relativistic)

5.4.1 Schrödinger equation with a central force

$$\nabla^2 u + \frac{2m}{\hbar^2}[E - U(r)]u = 0,$$

where m is the reduced mass of the electron and the nucleus,

$$m = m_e m_N/(m_e + m_N),$$

$$u = \sum_{lm} R_l(r)Y_{lm}(\theta, \phi)$$

and for a hydrogen-like atom with a Coulomb potential,

$$R_l''(r) + \frac{2}{r}R_l'(r) + \left[\frac{2m}{\hbar^2}\left(E + \frac{Z\alpha\hbar c}{r}\right) - \frac{l(l+1)}{r^2}\right]R_l(r) = 0,$$

where $\alpha = e^2/4\pi\epsilon_0\hbar c$ is the fine-structure constant. For $E < 0$ the only acceptable solutions require

$$E = -\frac{mc^2(Z\alpha)^2}{2n^2} = -R_\infty\hbar c\frac{m}{m_e}\left(\frac{Z}{n}\right)^2,$$

where n is an integer, $R_\infty = m_e c\alpha^2/2\hbar = 10\,973\,731.5\,\mathrm{m}^{-1}$ is the Rydberg constant, $R_\infty\hbar c = 2.179\,87\times10^{-18}\,\mathrm{J} = 13.6067\,\mathrm{eV}$.

Putting

$$x = 2r/an, \qquad a = \hbar/mc\alpha Z = m_e a_0/mZ,$$

where $a_0 = 0.529\,1772\times10^{-10}$ m is the Bohr radius, leads to

$$x^2 R_{nl}''(x) + 2x R_{nl}'(x) - \left[\frac{x^2}{4} - nx + l(l+1)\right]R_{nl}(x) = 0$$

with solutions expressible in terms of Laguerre polynomials $L_p(x)$ and the associated polynomials $L_p^s(x)$,

$$L_p(x) = \frac{e^x}{p!}\frac{d^p}{dx^p}(x^p e^{-x}), \quad L_p^s(x) = \frac{d^s}{dx^s}L_p(x),$$

$$xL_p^{s''}(x) + (s+1-x)L_p^{s'}(x) + (x-s)L_p^s(x) = 0,$$
$$L_0(x) = 1, \qquad L_1(x) = 1-x, \qquad L_2(x) = 1 - 2x + x^2/2,$$
$$L_3(x) = 1 - 3x + 3x^2/2 - x^3/6,$$
$$L_4 = 1 - 4x + 3x^2 - 2x^3/3 + x^4/24,$$

$$L_p^s(x) = (-1)^s p! \sum_{k=0}^{p-s}\frac{(-x)^k}{k!(s+k)!(p-s-k)!}.$$

The standardization of these functions is defined differently by different authors. $L_p(0) = 1$ is in agreement with Abramowitz and Stegun[6]; Bethe and Salpeter[7] use $\lim_{x\to\infty}(-x)^{-p}L_p(x) = 1$. The associated Laguerre polynomial $L_p^{(s)}(x)$ (degree $p-s$) should be distinguished from the generalized Laguerre polynomial, $L_p^{(s)}(x) = (-1)^s L_{p+s}^s(x)$.

The normalized radial wave functions are

$$R_{nl}(r) = -\sqrt{\frac{(n-l-1)!}{2n(n+l)!}}\left[\frac{2}{an}\right]^{3/2}(2\xi)^l L_{n+l}^{(2l+1)}(2\xi)e^{-\xi}, \qquad \xi = \frac{r}{na},$$

$$R_{10} = \frac{2}{\sqrt{a^3}}e^{-r/a}, \qquad\qquad R_{30} = \frac{2}{3\sqrt{3a^3}}\left(1 - \frac{2r}{3a} + \frac{2r^2}{27a^2}\right)e^{-r/3a},$$

$$R_{20} = \frac{1}{\sqrt{2a^3}}\left(1 - \frac{r}{2a}\right)e^{-r/2a}, \quad R_{31} = \frac{8}{27\sqrt{6a^5}}r\left(1 - \frac{r}{6a}\right)e^{-r/3a},$$

$$R_{21} = \frac{1}{2\sqrt{6a^5}}re^{-r/2a}, \qquad\qquad R_{32} = \frac{4}{81\sqrt{30a^7}}r^2e^{-r/3a}.$$

5.4.2 Molecular orbitals

$$u_{nlm}(r) = R_{nl}(r)Y_{lm}(\theta, \phi),$$

$$u(1s) = \frac{e^{-r/a}}{\sqrt{4\pi a^3}},$$

$$u(2s) = \frac{e^{-r/2a}}{2\sqrt{2\pi a^3}}(1 - r/2a),$$

$$u(2p) = \frac{e^{-r/2a}}{4\sqrt{2\pi a^5}}\begin{cases} x, & x = \sqrt{\frac{2\pi}{3}}r(Y_{1,-1} - Y_{1,+1}), \\ y, & y = i\sqrt{\frac{2\pi}{3}}r(Y_{1,-1} + Y_{1,+1}), \\ z, & z = \sqrt{\frac{4\pi}{3}}rY_{1,0}, \end{cases}$$

$$u(3s) = \frac{e^{-r/3a}}{3\sqrt{3\pi a^3}}(1 - 2r/3a + 2r^2/27a^2),$$

$$u(3p) = \frac{4e^{-r/3a}}{27\sqrt{2\pi a^5}}(1 - r/6a) \times \begin{cases} x \\ y \\ z \end{cases},$$

$$u(3d) = \frac{2e^{-r/3a}}{81\sqrt{2\pi a^7}} \times \begin{cases} [z^2 - \frac{1}{2}(x^2 + y^2)]/\sqrt{3} &= \sqrt{\frac{4\pi}{15}}r^2 Y_{2,0}, \\ xy &= i\sqrt{\frac{2\pi}{15}}r^2(Y_{2,-2} - Y_{2,+2}), \\ zx &= \sqrt{\frac{2\pi}{15}}r^2(Y_{2,-1} - Y_{2,+1}), \\ zy &= i\sqrt{\frac{2\pi}{15}}r^2(Y_{2,-1} + Y_{2,+1}), \\ (x^2 - y^2) &= \sqrt{\frac{2\pi}{15}}r^2(Y_{2,-2} + Y_{2,+2}). \end{cases}$$

5.5 Phase shift analysis

The wave equation for a spinless scalar interacting with a fixed finite-range scatterer,

$$\nabla^2\psi + k^2\psi = 0,$$

can be expressed as a superposition of incident and scattered waves,

$$\psi = \psi_{\text{inc}} + \psi_{\text{scat}}.$$

The incident wave may be described by a plane wave

$$\psi_{\text{inc}} = e^{ikz} = e^{ikr \cos \theta} = \sum_{l=0} A_l(kr) P_l(\cos \theta),$$

where

$$A_l(kr) = (2l + 1)i^l j_l(kr) = (2l + 1)i^l \sqrt{\frac{\pi}{2kr}} J_{l+1/2}(kr),$$

$$j_l(kr) \xrightarrow[r \to \infty]{} \frac{1}{kr} \sin(kr - \frac{\pi}{2} l),$$

$$e^{ikz} \to \frac{1}{kr} \sum_{l=0} (2l + 1)i^l P_l(\cos \theta) \sin(kr - \pi l/2),$$

$$\to \frac{1}{2ikr} \sum_{l=0} (2l + 1) P_l(\cos \theta)[e^{ikr} - i^l e^{-ikr}].$$

The scattering introduces a phase shift into the outgoing wave:

$$\psi_{\text{scat}} \to \frac{1}{2ikr} \sum_{l=0} (2l + 1) P_l(\cos \theta)(e^{-2\delta_l} - 1)e^{ikr}.$$

The differential scattering cross section is given by

$$\frac{d\sigma}{d\Omega} = \frac{v_{\text{scat}}}{v_{\text{inc}}} \lim_{r \to \infty} r^2 |\psi_{\text{scat}}(r, \theta)|^2.$$

The cross section for elastic scattering for a given angular momentum l is

$$\left(\frac{d\sigma}{d\Omega}\right)_l = |(2l + 1) P_l(\cos \theta)|^2 \sin^2 \delta_l$$

and the total cross section

$$\int \left(\frac{d\sigma}{d\Omega}\right) d\Omega = 4\pi(2l + 1) \sin^2 \delta.$$

5.6   Thermodynamics[8]

5.6.1   Zeroth law of thermodynamics

If system A is in thermal equilibrium with system T (thermal contact with no net exchange of heat) and system B is in thermal equilibrium with system T, then system A is in thermal equilibrium with system B. Temperature can be defined; a measure of the temperature is determined by the state (volume, length, electrical resistance, etc.) of system T (the thermometer).

Table 5.7 Thermodynamic systems and coordinates (variables).

System	Intensive variable (generalized force)		Extensive variable (generalized force)		$Work$ displacement
Chemical	Pressure	p	Volume	V	$p\,dV$
Surface	Surface tension	σ	Area	A	$-\sigma\,dA$
Elongation	Tension	$\mathcal{T}$	Length	l	$-\mathcal{T}\,dl$
Electric cell	emf	$\mathcal{E}$	Charge	q	$\mathcal{E}\,dq$
Electric field	Electric intensity	E	Polarization	P	$E{\cdot}dP$
Magnetic substance	Magnetic induction	B	Magnetization	M	$B{\cdot}dM$

5.6.2 Amount of substance

For most of chemistry, the number of molecules of a substance in a sample is of more importance than the mass of the substance in the sample. However, the number of molecules N is an inconvenient quantity for most situations, and the actual numbers of molecules are not as important as the relative amounts. In addition, the mass of substance is more readily determined. *Amount of substance* was introduced into SI as an independent base quantity in order to preserve the coherence of the system and to avoid the introduction of an ill-defined (and often unnecessary) conversion factor into physical relationships. (*See* Chapter 2.)

Amount of substance is a measure of the number of entities (atoms, molecules, ...) of a given substance; its *unit* is the *mole*. The *Avogadro constant*

$$N_{\mathrm{A}} = 6.022\,1367(36) \times 10^{23}\,\mathrm{mol}^{-1}$$

is the conversion factor from amount of substance n to a direct count of entities N.

5.6.3 Equation of state

An equation of state defines the relationship among the variables of the system. Most commonly, this is an expression relating the temperature, pressure, and volume of a specified amount of substance:

$$V = F(p, T) \quad \text{or} \quad f(p, V, T) = 0.$$

The work done against external pressure in a transformation from an initial state A to final state B is given by

$$w = \int_A^B p \, dV.$$

In an *isobaric* (also called *isopiestic*) transformation the pressure is constant; in an *isothermal* transformation the temperature is constant; in an *isochoric* transformation the volume of the system is constant and no work is done against external pressure.

Additional variables may be required to define the state of the system: amounts of substance of constituents (if more than a single substance is present), the surface and interfacial areas (if surface tension is important), electromagnetic field variables, etc.

5.6.4   Ideal gas law

The equation of state of an amount of substance n of an ideal gas is $pV = nRT$, where R is the universal gas constant, $R = 8.3145 \, \text{J mol}^{-1} \text{K}^{-1}$.

For an ideal gas, the work done against external pressure is

$$dw = pdV, \quad w = nRT \ln \frac{V_2}{V_1} = nRT \ln \frac{p_1}{p_2}.$$

5.6.5   First law

The internal energy U is a state function of the system. In a transformation of a system between states A and B the change in internal energy $\Delta U = U_B - U_A$ is

$$\Delta U = Q - w,$$

where Q is the amount of energy received by the system in forms other than (reversible) work. For an infinitesimal transformation

$$dU + dw = dQ.$$

5.6.6   Second law

Kelvin Postulate: A transformation whose only final result is to transform into work heat extracted from a source that is at the same temperature throughout is impossible.

Table 5.8 Thermodynamic coefficients.

Volume expansivity, coefficient of volume expansion	β	$\dfrac{1}{V}\dfrac{\partial V}{\partial T}\Big|_p$
Isothermal bulk modulus	B	$-V\dfrac{\partial p}{\partial V}\Big|_T$
Adiabatic bulk modulus	B_S	$-V\dfrac{\partial p}{\partial V}\Big|_S$
Isothermal compressibility	κ	$-\dfrac{1}{V}\dfrac{\partial V}{\partial p}\Big|_T$
Adiabatic compressibility	κ_S	$-\dfrac{1}{V}\dfrac{\partial V}{\partial p}\Big|_S$
Heat capacity at constant volume	C_v	$\dfrac{dQ}{dT}\Big|_V = T\dfrac{\partial S}{\partial T}\Big|_V$
Heat capacity at constant pressure	C_p	$\dfrac{dQ}{dT}\Big|_p = T\dfrac{\partial S}{\partial T}\Big|_p$
Heat capacity ratio	γ	C_p/C_v
Joule coefficient	η	$\dfrac{\partial T}{\partial V}\Big|_U$
Joule-Thomson (Kelvin) coefficient	μ	$\dfrac{\partial T}{\partial p}\Big|_H$
Linear expansivity	α	$\dfrac{1}{l}\dfrac{\partial l}{\partial T}\Big|_{\mathcal{T}}$ [a]
Isothermal Young's modulus	E	$\dfrac{1}{A}\dfrac{\partial \mathcal{T}}{\partial l}\Big|_T$ [a]
Adiabatic Young's modulus	E_S	$\dfrac{1}{A}\dfrac{\partial \mathcal{T}}{\partial l}\Big|_S$ [a]

[a] A = cross sectional area; $\mathcal{T}$ = tension.

Clausius Postulate: A transformation whose only result is to transfer heat from a body at a lower temperature to a body at a higher temperature is impossible.

Carnot Cycle: The Carnot heat engine cycles through four steps to convert heat to work: (1) isothermal expansion of a fluid at temperature T_2 from A to B with heat input, (2) adiabatic expansion from B to C, (3) isothermal compression at temperature T_1 from C to D with heat rejected to the surroundings, and (4) adiabatic compression from D back to A. The efficiency of the cycle, η, the ratio of the work done to the heat input at the higher temperature is

$$\eta \leq \frac{T_2 - T_1}{T_2}.$$

The maximum efficiency is achieved only when the cycle is carried out reversibly, e.g., the successive states in the cycle differ by infinitesimals from equilibrium states, and is the highest possible efficiency that an engine working between the temperatures T_1 and T_2 can have.

A Carnot cycle operating in the reverse sense can be used to extract an amount of heat Q_1 from a source at the low temperature T_1 by absorbing an amount of work w. Thus, if T_2 is the temperature of the receiver,

$$Q_1 < w\frac{T_1}{T_2 - T_1}.$$

An actual refrigerator will always have a lower efficiency because irreversible processes are always involved in a real device.

Entropy: In a reversible transformation

$$S(B) - S(A) = \int_A^B \frac{dQ}{T},$$

where $S(B)$ and $S(A)$ are the entropies of states B and A respectively; the integral is independent of the path over which the transformation proceeds. For an irreversible transformation

$$S(B) - S(A) > \int_A^B \frac{dQ}{T}.$$

Unlike dQ, dS is a perfect differential. Mathematically, temperature is defined by this condition: *Temperature is the integrating factor that makes dQ/T a perfect differential and S a thermodynamic state function.*

In differential form,

$$dS = \frac{dQ}{T}, \qquad dU = TdS - dw.$$

5.6.7 Third law

It is observed that the entropy change in an isothermal chemical reaction approaches zero as the temperature approaches absolute zero (Nernst's Law, 1905).

$$\lim_{T \to 0} \Delta S\big|_T = 0.$$

This has been generalized by Max Planck (1913) in the statement:

The entropy of a pure substance in a perfect crystalline form is zero at absolute zero.

Table 5.9 Derivatives of T, p, V, and S.   Only half of the 24 possible derivatives are listed here; the other 12 may be found by inversion: e.g.,

$$\frac{\partial p}{\partial V}\Big|_T = \left[\frac{\partial V}{\partial p}\Big|_T\right]^{-1}.$$

The Maxwell relations further constrain these derivatives.

$\dfrac{\partial V}{\partial p}\Big|_T = \kappa V$	$\dfrac{\partial p}{\partial T}\Big|_V = -\beta/\kappa$	$\dfrac{\partial T}{\partial V}\Big|_p = 1/\beta V$
$\dfrac{\partial p}{\partial T}\Big|_S = C_p/\beta VT$	$\dfrac{\partial T}{\partial S}\Big|_p = T/C_p$	$\dfrac{\partial S}{\partial p}\Big|_T = -\beta V$
$\dfrac{\partial T}{\partial S}\Big|_V = T/C_v$	$\dfrac{\partial S}{\partial V}\Big|_T = \beta/\kappa$	$\dfrac{\partial V}{\partial T}\Big|_S = -\kappa C_v/\beta T$
$\dfrac{\partial S}{\partial V}\Big|_p = C_p/\beta VT$	$\dfrac{\partial V}{\partial p}\Big|_S = -\kappa V/\gamma$	$\dfrac{\partial p}{\partial S}\Big|_V = \beta T/\kappa C_v$

5.6.8   Thermodynamic potentials

Internal Energy: $U(V, S)$ $dU = T dS - p dV$.

Enthalpy: (also known as "heat content" or "heat function.")

$$H(p, S) = U + pV, \qquad dH = T dS + V dp.$$

Free Energy: (also known as "Helmholtz free energy" or "work function;" also denoted by F):

$$A(V, T) = U - TS, \qquad dA = -S dT - p dV.$$

Thermodynamic Potential: (also known as "Gibbs (free) energy." In earlier literature, denoted by F where the Helmholtz free energy is denoted by A):

$$G(p, T) = U + pV - TS, \qquad dG = -S dT + V dp.$$

5.6.9   Maxwell's relations

Since dU, dH, dA, and dG are perfect differentials, certain relations follow from the cross second derivatives:

Table 5.10 Derivatives of the thermodynamic potentials.

Internal energy, U	Enthalpy, H	Free energy, A	Gibbs' function, G
$\dfrac{\partial U}{\partial T}\Big|_p = C_p - p\beta V$	$\dfrac{\partial H}{\partial T}\Big|_p = C_p$	$\dfrac{\partial A}{\partial T}\Big|_p = -S - \beta p V$	$\dfrac{\partial G}{\partial T}\Big|_p = -S$
$\dfrac{\partial U}{\partial T}\Big|_V = C_v$	$\dfrac{\partial H}{\partial T}\Big|_V = C_v - \beta V/\kappa$	$\dfrac{\partial A}{\partial T}\Big|_V = -S$	$\dfrac{\partial G}{\partial T}\Big|_V = -S + \beta V/\kappa$
$\dfrac{\partial U}{\partial T}\Big|_S = \kappa p C_v/\beta T$	$\dfrac{\partial H}{\partial T}\Big|_S = C_p/\beta T$	$\dfrac{\partial A}{\partial T}\Big|_S = -S + p C_v \kappa/\beta T$	$\dfrac{\partial G}{\partial T}\Big|_S = -S + C_p/\beta T$
$\dfrac{\partial U}{\partial p}\Big|_T = V(\kappa p - \beta T)$	$\dfrac{\partial H}{\partial p}\Big|_T = V(1 - \beta T)$	$\dfrac{\partial A}{\partial p}\Big|_T = p\kappa V$	$\dfrac{\partial G}{\partial p}\Big|_T = V$
$\dfrac{\partial U}{\partial p}\Big|_V = \kappa C_v/\beta$	$\dfrac{\partial H}{\partial p}\Big|_V = V + \kappa C_v/\beta$	$\dfrac{\partial A}{\partial p}\Big|_V = -\kappa S/\beta$	$\dfrac{\partial G}{\partial p}\Big|_V = V - \kappa S\beta$
$\dfrac{\partial U}{\partial p}\Big|_S = \kappa p V/\gamma$	$\dfrac{\partial H}{\partial p}\Big|_S = V$	$\dfrac{\partial A}{\partial p}\Big|_S = V(\kappa p/\gamma - \beta T S/C_v)$	$\dfrac{\partial G}{\partial p}\Big|_S = V - \beta V T S/C_p$
$\dfrac{\partial U}{\partial V}\Big|_T = \beta T/\kappa - p$	$\dfrac{\partial H}{\partial V}\Big|_T = (\beta T - 1)/\kappa$	$\dfrac{\partial A}{\partial V}\Big|_T = -p$	$\dfrac{\partial G}{\partial V}\Big|_T = -1/\kappa$
$\dfrac{\partial U}{\partial V}\Big|_p = C_p/\beta V$	$\dfrac{\partial H}{\partial V}\Big|_p = C_p/\beta V$	$\dfrac{\partial A}{\partial V}\Big|_p = -p - S/\beta V$	$\dfrac{\partial G}{\partial V}\Big|_p = -S/\beta V$
$\dfrac{\partial U}{\partial V}\Big|_S = -p$	$\dfrac{\partial H}{\partial V}\Big|_S = -\gamma/\kappa$	$\dfrac{\partial A}{\partial V}\Big|_S = S\beta T/\kappa C_v - p$	$\dfrac{\partial G}{\partial V}\Big|_S = (-\gamma + S\beta T/C_v)/\kappa$
$\dfrac{\partial U}{\partial S}\Big|_T = T - \kappa p/\beta$	$\dfrac{\partial H}{\partial S}\Big|_T = T - 1/\beta$	$\dfrac{\partial A}{\partial S}\Big|_T = -\kappa p/\beta$	$\dfrac{\partial G}{\partial S}\Big|_T = 1/\beta$
$\dfrac{\partial U}{\partial S}\Big|_p = T - \beta p V T/C_p$	$\dfrac{\partial H}{\partial S}\Big|_p = T$	$\dfrac{\partial A}{\partial S}\Big|_p = -(S - p V\beta)T/C_p$	$\dfrac{\partial G}{\partial S}\Big|_p = -TS/C_p$
$\dfrac{\partial U}{\partial S}\Big|_V = T$	$\dfrac{\partial H}{\partial S}\Big|_V = T + (\gamma - 1)/T$	$\dfrac{\partial A}{\partial S}\Big|_V = -TS/C_p$	$\dfrac{\partial G}{\partial S}\Big|_V = -T(S - \beta V/\kappa)/C_v$

$$\frac{\partial^2 U}{\partial V \partial S}: \quad -\left.\frac{\partial T}{\partial V}\right|_S = \left.\frac{\partial p}{\partial S}\right|_V = -\beta T/\kappa C_v,$$

$$\frac{\partial^2 H}{\partial p \partial S}: \quad \left.\frac{\partial T}{\partial p}\right|_S = \left.\frac{\partial V}{\partial S}\right|_p = \beta V T/C_p,$$

$$\frac{\partial^2 A}{\partial V \partial T}: \quad \left.\frac{\partial S}{\partial V}\right|_T = \left.\frac{\partial p}{\partial T}\right|_V = \beta/\kappa,$$

$$\frac{\partial^2 G}{\partial p \partial T}: \quad -\left.\frac{\partial S}{\partial p}\right|_T = \left.\frac{\partial V}{\partial T}\right|_p = \beta V.$$

5.6.10   Heat capacity

At constant volume:

$$C_v \equiv \left.\frac{dQ}{dT}\right|_V = \left.\frac{\partial U}{\partial T}\right|_V.$$

At constant pressure:

$$C_p = \left.\frac{dQ}{dT}\right|p = \left.\frac{\partial H}{\partial T}\right|_p = \left.\frac{\partial U}{\partial T}\right|_p + p\left.\frac{\partial V}{\partial T}\right|_p,$$

$$C_p - C_v = \left[p + \left.\frac{\partial U}{\partial V}\right|_T\right]\left.\frac{\partial V}{\partial T}\right|_p = \frac{V\beta^2 T}{\kappa}.$$

An ideal gas is defined by $\left.\dfrac{\partial U}{\partial V}\right|_T = 0$ and $C_p - C_v = R$. From the kinetic theory of ideal gases, $C_v = 3R/2$ for a monatomic gas and $5R/2$ for a diatomic gas; then

$$\gamma \equiv C_p/C_v = \begin{cases} 5/3, & \text{monatomic gas}, \\ 7/5, & \text{diatomic gas}. \end{cases}$$

For an ideal gas:

$$S = C_v \ln T + R \ln V + S^\circ,$$

where S° is a constant of integration, and is the entropy of a "standard state," e.g., the entropy of the gas at standard conditions of temperature and pressure.

In an adiabatic transformation, the system is thermally isolated; no heat is exchanged with the environment and the transformation is reversible. For an ideal gas

$$pV^\gamma = \text{const}, \qquad TV^{\gamma-1} = \text{const}.$$

5.6.11 Clapeyron equation

Along the equilibrium curve between two phases of a system, the free energy, $G = U + pV - TS$ is a constant. Since for a reversible process, $dG = V\,dp - S\,dT$, and is equal in the two phases if they are in equilibrium, it follows that $(V_2 - V_1)dp = (S_2 - S_1)dT$ where 1 and 2 refer to the variables in the first and second phases, respectively. From this, since $(S_2 - S_1)T = L_{12}$, the latent heat of transformation from phase 1 to phase 2, the change in pressure along the phase transition curve is

$$\frac{dp}{dT} = \frac{L_{12}}{(V_2 - V_1)T}.$$

If phase 2 is a vapor phase at low density so that $V_2 \gg V_1$ and the ideal gas law is a good approximation, and if L_{12} is approximately constant, the vapor pressure of a solid or liquid is

$$p = p^* e^{-L_{12}/RT} \qquad \text{(Clausius-Claperon equation)}.$$

The Clapeyron equation applies also to the melting of a solid phase ($V_1 = V_s$, $V_2 = V_l$). Since the volume change on melting is small, the change in the melting point with pressure is small. If the material expands on melting ($V_s < V_l$, as for most materials), $dT/dp > 0$ and an increase in pressure increases the melting point; for water, $V_l < V_s$ and increased pressure decreases the melting point.

5.6.12 Basic thermodynamic relations

$$C_v = \left.\frac{\partial U}{\partial T}\right|_V = T\left.\frac{\partial S}{\partial T}\right|_V = -T\left.\frac{\partial p}{\partial T}\right|_V \left.\frac{\partial V}{\partial T}\right|_S,$$

$$C_p = \left.\frac{\partial H}{\partial T}\right|_p = T\left.\frac{\partial S}{\partial T}\right|_p = T\left.\frac{\partial V}{\partial T}\right|_p \left.\frac{\partial p}{\partial T}\right|_S,$$

$$C_p - C_v = T\left.\frac{\partial p}{\partial T}\right|_V \left.\frac{\partial V}{\partial T}\right|_p = -T\left.\frac{\partial V}{\partial T}\right|_p^2 \left.\frac{\partial p}{\partial V}\right|_T = \frac{\beta^2 VT}{\kappa},$$

$$\gamma = \frac{C_p}{C_v} = \frac{\left.\frac{\partial p}{\partial V}\right|_S}{\left.\frac{\partial p}{\partial V}\right|_T} = \frac{\kappa}{\kappa_S},$$

$$\left.\frac{\partial C_v}{\partial V}\right|_T = T\left.\frac{\partial^2 p}{\partial T^2}\right|_V, \qquad \left.\frac{\partial C_p}{\partial p}\right|_T = -T\left.\frac{\partial^2 V}{\partial T^2}\right|_p,$$

$$TdS = C_p dT - T \frac{\partial V}{\partial T}\Big|_p dp = C_p dT - V\beta T dp,$$

$$= C_v \frac{\partial T}{\partial p}\Big|_V dp + C_p \frac{\partial T}{\partial V}\Big|_p dV = \frac{1}{\beta}\left[\kappa C_v dp + \frac{C_p}{V}dV\right]$$

$$= C_v dT + T \frac{\partial p}{\partial T}\Big|_V dV = C_v dT + \frac{\beta T}{\kappa}dV.$$

Joule Coefficient:

$$\eta = \frac{\partial T}{\partial V}\Big|_U = -\frac{1}{C_v}\left[T\frac{\partial p}{\partial T}\Big|_V - p\right] = -\frac{1}{C_v}\left[\frac{\beta T}{\kappa} - p\right].$$

Joule-Thomson (Kelvin) coefficient:

$$\mu = \frac{\partial T}{\partial p}\Big|_H = \frac{1}{C_p}\left[T\frac{\partial V}{\partial T}\Big|_p - V\right] = \frac{V}{C_p}(\beta T - 1).$$

References

[1] I. Newton, *Mathematical Principles of Natural Philosophy*, translated by A. Motte, revised by Florian Cajori (University of California Press, Berkeley, 1936).

[2] J. Korteweg and G. deVries, Philos. Mag. **39**, 422 (1895); N.J. Zabrusky and M. D. Kruskal, Phys. Rev. Lett. **15**, 240 (1965).

[3] J. D. Jackson, *Classical Electrodynamics* (Wiley, New York, 1975).

[4] David L. Book, *NRL Plasma Formulary* (Naval Research Laboratory, Washington, D.C., 1980).

[5] P. M. Morse and H. Feshbach, *Methods of Theoretical Physics*, (McGraw-Hill, New York, 1975).

[6] *Handbook of Mathematical Functions*, edited by M. Abramowitz and I. Stegun, NBS AMS-55 (U.S. Government Printing Office, Washington, D.C., 1964).

[7] H. A. Bethe and E. E. Salpeter, *Quantum Mechanics of One- and Two-Electron Atoms* (Academic, New York, 1957).

[8] R. A. Alberty and R. J. Silbey, *Physical Chemistry* (Wiley, New York, 1992).

[9] *Reference Data for Radio Engineers* (SAMS, Indianapolis, 1977).

6

ENGINEERING PHYSICS

6.1 Engineering elasticity

6.1.1 Elastic constants

In engineering applications involving homogeneous, isotropic materials the most common constants are the *bulk modulus* κ, the *shear modulus* μ, *Young's modulus* E, and *Poisson's ratio* ν. The force per unit area on a surface element within a body may be either a normal (perpendicular) stress (σ) or a shear (tangential) stress (τ).

Bulk Modulus: The ratio of the normal force per unit area of the surface (pressure) to the relative change in volume:

$$\kappa = -\frac{F_{\mathrm{n}}/A}{\Delta V/V} \qquad (F_{\mathrm{n}}/A = p = \text{pressure}, \quad V = \text{volume}).$$

Modulus of Elasticity (Young's modulus): A rectangular bar of length l_1 will stretch under tension an amount e_1; at the same time, its transverse dimensions l_2, l_3 will contract amounts e_2 and e_3. The normal strains are

$$\epsilon_1 = e_1/l_1, \qquad \epsilon_2 = \epsilon_3 = -e_3/l_3 = -\epsilon_2/l_2.$$

Poisson's Ratio: $\quad \nu = \epsilon_2/\epsilon_1,$

$$E = \frac{F_{\mathrm{T}}/A}{e_1/l_1} = \frac{\sigma}{\epsilon_1} \qquad (F_{\mathrm{T}} = \text{tensile force}, \quad A = \text{area}).$$

The forces of shear act tangentially to the surface and cause a relative slippage δ of parallel surfaces separated by a distance l. The shear strain is the displacement angle γ, or (assuming small deformations) the ratio of the displacement to the separation distance

$$\gamma = \delta/l.$$

Shear Modulus:

$$G = \frac{F_s/A}{\delta/l} = \frac{\tau}{\gamma},$$

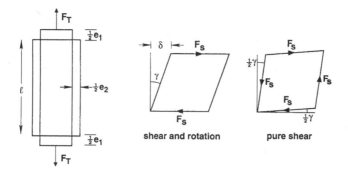

shear and rotation   pure shear

tensile strain $\epsilon_1 = e_1/l_1$,   shearing strain $\gamma = \delta/l$.

Since the elastic properties of an isotropic medium are expressible in terms of two independent quantities, these constants are related (see Section 5.1.5) by

$$E = \frac{9\kappa\mu}{3\kappa + \mu} = 3\kappa(1 - 2\nu) = 2\mu(1 + \nu), \qquad \nu = \frac{3\kappa - 2\mu}{2(3\kappa + \mu)}.$$

Table 6.1 Elastic constants of solids. The elastic constants of a material depend to a great extent on its crystalline structure, past history, etc. The values below can only be taken as approximate.

Material	Bulk modulus $\dfrac{\kappa}{(10^{10}\,\text{Pa})}$	Young's modulus $\dfrac{E}{(10^{10}\,\text{Pa})}$	Shear modulus $\dfrac{\mu}{(10^{10}\,\text{Pa})}$	Poisson's ratio ν
Aluminum	7.5	7	2.5	0.34
Al alloys				
Al-Si-Mg		6.8-7.2	2.6-2.8	0.34
Al-Cu-Mg		7.2-7.4	2.8	0.34
Bismuth	3	3.2	1.2	0.33
Cadmium	4.2	5	2	0.3
Copper	13.5	11	4.4	0.34
Glass, crown	5	6	2.5	0.25
Gold	16.5	8	2.8	0.42
Iron				
Wrought	16	21	7.7	0.28
Cast	9.5	11	5	0.27
Lead	4.1	1.6	0.6	0.44
Magnesium	3.3	4.1	1.7	0.3
Nickel	17	21	7.8	0.3
Platinum	24.5	17	6.3	0.39
Silver	10.5	7.7	2.8	0.37
Steel				
Cast	17	20	7.5	0.28
Mild	16	22	8.0	0.28
$\sim 15\%$ Ni		20.3-20.5	7.8	0.31
$\sim 30\%$ Ni		18.2	7	0.3
$>35\%$ Ni		15.6	6	0.3
Tin	5.3	5.3	1.9	0.33
Tungsten	30	39	15	0.35
Zinc	3.5	8	3.6	0.23
Brass	6	9	3.5	0.35
Phosphor bronze		12	4.4	0.38
German silver	15	11	4.5	0.37

6.1.2 Deflections in uniform beams

The approximation required for the simplified elastic description of a beam is that the cross sectional dimensions are small compared to the radius of curvature of bending. Consequently, the elastic displacements across the beam may be considered to vary linearly, and plane cross sections remain plane. At any point x along the beam the deflection y is governed by the differential equations

$$EI\frac{d^2y}{dx^2} = M, \qquad EI\frac{d^4y}{dx^4} = w + \sum F_i\delta(x - x_i),$$

where E is Young's modulus, I is the geometric moment of inertia of the cross sectional area,

$$I \equiv \int\int [(\xi - \bar{\xi})^2 + (\eta - \bar{\eta})^2]\,d\xi d\eta,$$

M is the bending moment, w is the beam loading due to gravitational forces (weight per unit length) plus any other distributed loads, and F_i are external point forces imposed at x_i.

In a cantilevered uniform beam of length L with a perpendicular force F at one end, neglecting the weight of the beam itself the cross section at a distance x from F has a bending moment $M = Fx$.

Across the section the fiber stress increases linearly with distance from the neutral axis. At a distance ξ from the neutral axis the stress is given by

$$\sigma = M\xi/I.$$

The maximum fiber stress at any section occurs at a point $\xi = c$ most remote from the neutral axis. Hence

$$\sigma_b = M_b/(I/c),$$

where I/c is the *section modulus* of the beam. The maximum fiber stress in the beam occurs at the section of greatest bending moment. In designing structures, care must be taken to insure that the maximum fiber stress is within the elastic range of the material.

The shearing force is related to the bending moment by $F_s = dM/dx$. The maximum shear stress τ_b occurs at the section of greatest vertical shear. The maximum shear stress is $\tau_b = \beta F_s/A$ where F_s/A is the *average* shear on the section and β is a factor that depends on the shape of the section. For a rectangular section $\beta = 3/2$; for a solid circular cross section $\beta = 4/3$.

Horizontal cantilevered beam: Uniform beam, length L; deflection due to weight of beam w per unit length, and a force F applied at distance z_1 from the fixed end:

$$\delta(z) = \frac{wz^2}{24EI}(6L^2 - 4Lz + z^2) + \frac{F}{6EI}g(z, z_1), \qquad 0 < z < L,$$

where

$$g(z, z_1) = \begin{cases} (3z_1 - z)z^2, & 0 \le z \le z_1 \\ (3z - z_1)z_1^2, & z_1 \le z \le L \end{cases}$$

$$\delta(L) = \frac{w}{8EI}L^4 + \frac{F}{6EI}L^2(3L - z_1).$$

Uniform Horizontal Beam Clamped at Both Ends: Length L, $0 < z < L$;
1. Deflection due to weight of beam, w per unit length:

$$\delta(z) = \frac{w}{24EI}z^2(L - z)^2, \qquad \delta_{\max} = \delta(0) = \frac{wL^4}{384EI}.$$

2. Deflection due to concentrated load at z_1, $F \gg wL$:

$$\delta(z) = \frac{F}{6EIL^3}[L(z + z_1 + 2|z - z_1|) - 2zz_1]g_2(z, z_1, L),$$

$$g_2(z, z_1, L) = \begin{cases} z^2(L - z_1)^2, & 0 < z < z_1 \\ (L - z)^2 z_1^2, & z_1 < z < L. \end{cases}$$

The maximum deflection is

$$\delta(y_{\mathrm{m}}) = \frac{F}{3EI}z_1^2(L - z_1)^2 \frac{(L + |L - 2z_1|)}{(2L + |L - 2z_1|)},$$

$$y_{\mathrm{m}} = \frac{L}{2}\left[1 - \frac{L - 2x_1}{2L + |L - 2x_1|}\right].$$

For the load at the center of the beam, $z_1 = L/2$, the maximum deflection is

$$y(L/2) = FL^3/192EI.$$

Symmetrically Supported Beam: Uniform beam, length L, supported in a horizontal plane at points $+aL/2$ and $-aL/2$ from the center ($0 < a < 1$), deflection due to weight of beam

$$\delta(xL/2) = \frac{wL^4}{1536EI}\left[x^2(x^2 + 6) - \begin{cases} 4a(3x^2 + a^2) & 0 < |x| < a \\ 4|x|(3a^2 + x^2) & a < |x| < 1 \end{cases}\right].$$

1. *Beam supported at ends:* , $a = 1$
   maximum deflection, $\delta(0) = 5wL^4/384EI$.

2. *Beam horizontal at ends:* , $y'(1) = y'(-1) = 0$, $a = 1/\sqrt{3}$,
   maximum deflection, $\delta(0) = \dfrac{wL^4}{16EI}\dfrac{9 - 4\sqrt{3}}{216} = \dfrac{wL^4}{1668.12EI}$.

3. *Minimal maximum deflection:* , $y(1) = y(0) = y(-1)$
   $$4a^3 - 12a^2 + 3 = 0 \quad a = 0.5537,$$
   maximum deflection, $\delta(0) = \dfrac{wL^4}{3698.29EI}$.

6.2   Friction

The coefficient of *static friction* is the ratio of the force F required to initiate sliding between two surfaces to the normal force N that presses them together: $f = F/N$. The coefficient of *sliding friction* is the same ratio for the force required to maintain constant motion once it has been initiated. The *angle of repose* is the angle θ with respect to the horizontal of a surface for which sliding will just begin: $\tan \theta = f$.

Belt friction opposes the slipping of a belt on a pulley. When power is transmitted, the tension T_1 on the driving side is greater than the tension T_2 on the driven side. Neglecting centrifugal force,

$$T_1 = T_2 e^{f\alpha},$$

where α is the angular extent of the contact between the pulley and the belt. If v is the speed of the pulley surface, the power transmitted is $P = (T_1 - T_2)v$.

Greasy Lubrication: The lubricant is thin but adheres strongly enough to the rubbing surfaces that the layers of lubricant slip over one another.

Viscous Lubrication: The lubricant is thick enough to carry the entire hydrostatic or hydrodynamic pressure.

Table 6.2 Coefficients of static and sliding friction[4].

Materials, lubricant	Static	Sliding
Hard steel / hard steel		
Dry	0.78	0.42
Oleic acid	0.11	0.12
Light mineral oil	0.23	
Castor oil	0.15	0.081
Stearic acid	0.0052	0.029
Hard steel / graphite		
Dry	0.21	
Oleic acid	0.09	
Mild steel / lead		
Dry	0.95	0.95
Medium mineral oil	0.5	0.3
Aluminum / aluminum, dry	1.05	1.4
Aluminum / mild steel, dry	0.61	0.47
Glass / glass		
Dry	0.94	0.40
Ricinoleic acid	0.005	
Oleic acid		0.09
Teflon / teflon, dry	0.04	

Table 6.2 *Continued.*

Materials, lubricant	Static	Sliding
Teflon / steel, dry	0.04	

6.3 Electromagnetic frequency bands

Table 6.3 Electromagnetic frequency/wavelength bands.
Nomenclature: F, frequency; H, high; M, medium; L, low; V, very;
E, extreme; S, super; U, ultra.

Designation	Frequency range		Wavelength range	
ULF	< 100	Hz	> 3	Mm
ELF	100—3000	Hz	100—3000	km
VLF	3— 30	kHz	10— 100	km
LF	30— 300	kHz	1— 10	km
MF	300—3000	kHz	100—1000	km
HF	3— 30	MHz	10— 100	m
VHF	30— 300	MHz	1— 10	m
UHF	300—3000	MHz	10— 100	cm
SHF (microwave)	3— 30	GHz	1— 10	cm
EHF	30— 300	GHz	1— 10	mm
Submillimeter	300—3000	GHz	100—1000	μm
Infrared	> 3	THz	< 100	μm

Microwave subbands				
S	2.6 — 3.95	GHz	7.6 —11.5	cm
G	3.95— 5.85		5.1 — 7.6	
J	5.3 — 8.2		3.7 — 5.7	
H	7.02—10.0		3.0 — 4.25	
X	8.2 —12.4		2.4 — 3.7	
M	10.0 —15.0		2.0 — 3.0	
P	12.4 —18.0		1.67— 2.4	
K	18.0 —26.5		1.1 — 1.67	
R	26.5 —40.0		0.75— 1.1	

NEW BAND DESIGNATIONS

A	— 250 MHz	F	3— 4 GHz	K	20— 40 GHz
B	250— 500 MHz	G	4— 6 GHz	L	40— 60 GHz
C	500—1000 MHz	H	6— 8 GHz	M	60—140 GHz
D	1— 2 GHz	I	8—10 GHz		
E	2— 3 GHz	J	10—20 GHz		

Table 6.4 Accepted US units for magnetic properties[7]. Gaussian electromagnetic units (cgs electromagnetic units) are accepted in the United States along with the coherent SI units. The defining relation for Gaussian magnetic quantities is $B = H + 4\pi M$ while the SI units are based on the definitions $B = \mu_0(H + M)$, but the quantity $B - \mu_0 H$ is often called "magnetic polarization" and denoted by J or I. In the Gaussian system $\mu_0 = 1$, both oersted and gauss have dimension $cm^{-1/2} \cdot g^{1/2} \cdot s^{-1}$ and a conversion factor relating these units to tesla $(kg \cdot s^{-2} \cdot A^{-1})$ is not strictly appropriate.

Quantity	Symbol	SI unit	Gaussian unit[a]	Effective conversion factor[b]
Magnetic flux density, magnetic induction	B	T, Wb/m^2	G	10^{-4} T/G
Magnetic flux	Φ	Wb, V/s	Mx	10^{-8} Wb/Mx
Magnetic potential, magnetomotive force	U, F	A	Gb	$(10/4\pi)$ A/Gb
Magnetic field strength, magnetizing force	H	A/m	Oe, Gb/cm	$(10^3\,4\pi)$ A·m^{-1}/Oe
(Volume) magnetization[c]	M	A/m	emu/cm^3 [d]	10^3 A·m^{-1}/G
Magnetic polarization, (Volume) magnetization	$4\pi M$	A/m	G	$(10^3/4\pi)$ A·m^{-1}/G
intensity of magnetization	J, I	T, Wb/m^2	emu/cm [d]	$4\pi \times 10^{-4}$ T$/$(emu/cm^3) [e]
(Mass) magnetization	σ, M	$A \cdot m^2/kg$	emu/g [d]	1 (A·m^2/kg)$/$(emu/g)
		$Wb \cdot m/kg$		$4\pi \times 10^{-7}$ (Wb·m/kg)$/$(emu/g)
Magnetic moment	m	$A \cdot m^2$, J/T	erg/G	10^{-3} (J/T)$/$(erg/G)
Magnetic dipole moment	j	Wb/m	erg/G	$4\pi \times 10^{-10}$ (Wb/m)$/$(erg/G) [d]
(Volume) susceptibility	χ, κ	1	1	4π
		$Wb/(A \cdot m)$		$(4\pi)^2 \times 10^{-7}$ H/m
(Mass) susceptibility	χ_ρ, κ_ρ	m^3/kg	cm^3/g	$4\pi \times 10^{-3}$ (m^3/kg)$/$(cm^3/g)
		$H \cdot m^2/kg$		$(4\pi)^2 \times 10^{-10}$
Permeability	μ	H/m	1	$4\pi \times 10^{-7}$ H/m
(Volume) energy density, energy product[f]	W	J/m^3	erg/cm^3	0.1 (J/m^3)$/$(erg/cm^3)
Demagnetization factor	D, N	1	1	$1/4\pi$

a Coherent cgs electromagnetic units: G, gauss; Mx, maxwell; Gb, gilbert; Oe, oersted.
b When two factors are given, the first is consistent with SI; the second (lower) one is not, and is based on the definition $B = \mu_0 H + J$.
c Magnetic moment per unit volume.
d "emu" (= electromagnetic unit) is not itself a unit but a designation of a unnamed coherent unit in the Gaussian system.
e Recognized under SI even though this is based on the definition $B = \mu_0 H + J$. See footnote a.
f In SI $B \cdot H$ and $\mu_0 M \cdot H$ have the unit J/m^3; in the Gaussian system $M \cdot H$ and $B \cdot H/4\pi$ have the unit erg/cm^3.

6.4 International temperature scale

The 1990 International Temperature Scale (ITS-90) provides defining fixed points intended to realize thermodynamic temperature from 0.65 K to 1357.77 K; above 1234.93 K the temperature scale is defined by the Planck radiation law and the radiance of a black body. It supersedes the previous International Practical Temperature Scale, IPTS-68, (amended edition, 1975) and the 1976 Provisional Temperature Scale (EPT-76) (0.5 K – 30 K).

1. From 0.65 K to 5 K ITS-90 is defined by the vapor pressure–temperature relation for liquid He:

^{3}He: between 0.65 K and 3.2 K (116 Pa to 107 kPa),

^{4}He II: between 1.25 K and the λ-point (2.1768 K) (115 Pa to 5041.8 Pa), and

^{4}He I: between the λ point and 5.0 K (5041.8 Pa to 196 kPa).

T_{90} is given by equations of the form

$$T_{90}/\text{K} = A_0 + \sum_{i=1} A_i \big[\big(\ln(p/\text{Pa}) - B\big)/C\big]^i.$$

2. From 3.0 K to 24.5561 (TPNe) ITS-90 is defined in terms of the He constant volume gas thermometer (CVGT), using a quadratic equation,

$$T_{90} = a + bp + cp^2,$$

fitted to three temperature points: a temperature between 3.0 K and 5.0 K determined using a ^{3}He or a ^{4}He I vapor pressure thermometer, and the triple points of equilibrium hydrogen (13.8033 K) and neon (24.5561 K). When a ^{3}He or ^{4}He I constant volume gas thermometer is used, the second virial coefficient $B_3(T_{90})$ or $B_4(T_{90})$ must be included to account for the non-ideal behavior of the gas.

3. From 13.8033 K (TPH) and 1234.93 K (FPAg) ITS-90 is defined in terms of the resistance ratios, $W(T_{90}) = R(T_{90})/R(273.16\,\text{K})$, of a standard platinum resistance thermometer (SPRT) in terms of reference function $W_r(T_{90})$ and a deviation function $W(T_{90}) - W_r(T_{90})$ determined by fitting to a number of fixed points in the range of application.

4. Above 1234.93 K, ITS-90 is defined by radiation thermometry based on Planck's law:

$$\frac{L_\lambda(T_{90})}{L_\lambda(T_{90}[X])} = \frac{\exp(c_2/\lambda T_{90}[X]) - 1}{\exp(c_2/\lambda T_{90}) - 1},$$

Table 6.5 ITS-90 Defining points.

vp, vapor pressure point;   tp, triple point;   fp, freezing point;   mp, melting point.

	Temperature		System	Symbol
	K	°C		
1	3 to 5	−270.15 to −268.15	Helium vp	
	4.2221	−268.9279		a
2	13.8033	−259.3467	e-Hydrogen tp	TPH
3	∼ 17	∼ −256.15	e-Hydrogen vp	b
4	∼ 20.3	∼ −252.85	e-Hydrogen vp	c
5	24.5561	−248.5939	Neon tp	TPNe
6	54.3584	−218.7918	Oxygen tp	TPO
7	83.8058	−189.3442	Argon tp	TPAr
8	234.3156	−38.8344	Mercury tp	TPHg
9	273.16	0.0100	Water tp	TPW
10	302.9146	29.7646	Gallium mp	MPGa
11	429.7485	156.5985	Indium fp	FPIn
12	505.078	231.928	Tin fp	FPSn
13	692.677	419.527	Zinc fp	FPZn
14	933.473	660.323	Aluminum fp	FPAl
15	1234.93	961.78	Silver fp	FPAg
16	1337.33	1064.18	Gold fp	FPAu
17	1357.77	1084.62	Copper fp	FPCu

[a] Boiling point of ^{4}He at a pressure of 101.325 kPa; not a defining point of ITS-90.
[b] Liquid/vapor equilibrium of e-H$_2$; for $p = 33.3306$ kPa: 17.0357 K. (e-H$_2$ is the equilibrium mixture of *ortho-* and *para-* molecular forms.)
[c] Boiling pt of e-H$_2$ at $p = 101.325$ kPa: 20.2711 K.

in which $L_\lambda(T_{90})$ and $L_\lambda(T_{90}[X])$ are the spectral luminosities of the radiance of a black body at vacuum wavelength λ at T_{90} and at $T_{90}[X]$, respectively, X referring to the freezing points of Ag ($T_{90}[\mathrm{Ag}] = 1234.93$ K), Au ($T_{90}[\mathrm{Au}] = 1337.33$ K), or Cu ($T_{90}[\mathrm{Cu}] = 1357.77$ K).

In ranges where overlapping definitions exist, they are considered to be equivalent; although numerical differences in interpolated values of T_{90} may be detectable in measurements of the highest precision, these differences are of negligible practical importance.

6.5   Heat transfer

6.5.1   Conduction

The flow of heat across an area A in a material with a temperature gradient dT/dx and thermal conductivity κ is given by

$$\Phi = -\kappa A\nabla T.$$

For a rectangular parallelepiped with opposite faces at T_1 and T_2 spaced x apart,

$$\Phi = -\kappa A(T_2 - T_1).$$

For two concentric isothermal cylindrical surfaces of radii r_1 and r_2 maintained at temperatures T_1 and T_2, respectively, and with length $L \gg r$,

$$\Phi = -\frac{2\pi L}{\ln(r_2/r_1)}\kappa(T_2 - T_1).$$

For two concentric spheres,

$$\Phi = -\frac{4\pi r_1 r_2}{r_2 - r_1}\kappa(T_2 - T_1).$$

The *time-dependent* differential equation for heat conduction is

$$\rho c\frac{\partial T}{\partial t} = \kappa\nabla^2 T,$$

where ρ is the density and c is the specific heat of the material.

Cooling (or heating) of slabs: Consider a slab of homogeneous material of thickness $2a$, small in comparison to its other dimensions. The faces of the slab coincide with the planes $x = +a$ and $x = -a$. The slab is initially at temperature T_o and is immersed in an environment at a fixed uniform temperature T_1. If the boundary condition is given by Newton's law of cooling the heat flow from the slab per unit area of the surface is given by

$$W = h(T(\pm a) - T_1)\qquad \text{for air } h \approx 11.7 \text{ W·m}^{-2}\text{·K}^{-1}$$

For slow cooling ($ha/\kappa \ll 1$), after a time larger than the thermal *relaxation time*, defined by $\tau = (2a/\pi)^2/D$, the temperature is approximately uniform spatially as the slab cools and is given by

$$T(x,t) = T_1 + (T_o - T_1)\left[1 + \frac{h}{6\kappa a}(a^2 - 3x^2)\right]e^{-hDt/\kappa a}.$$

The *thermal diffusivity* $D = \kappa/\rho c$ (or D_td) is often represented by h^2 in older literature.

Cylinders: For a cylinder of radius r_o and with $hr_o/\kappa \ll 1$, the temperature will be approximately uniform throughout the cylinder:

$$T(r,t) = T_1 + (T_o - T_1)\left[1 + \frac{h}{4\kappa r_o}(r_o^2 - 2r^2)\right]e^{-2hDt/\kappa r_o}.$$

Spheres: For a sphere of radius R the corresponding expression is

$$T(r,t) = T_1 + (T_o - T_1)\left[1 + \frac{h}{10\kappa R}(3R^2 - 5r^2)\right]e^{-3hDt/\kappa R}.$$

Table 6.6 Thermal conductivity and thermal diffusivity.
Materials are at 25 °C (298.15 K) unless otherwise specified.

Material	Conductivity κ	Diffusivity D
	W/m·K	$10^{-4}\,\mathrm{m^2/s}$
Aluminum[a]	237	0.969
Bismuth (polycrystalline)[a]	7.9	0.065
Brass (yellow)[b]	105	0.35
Chromium[a]	93.9	0.291
Constantin (60Cu,40Ni)[c]	24.3	0.069
Copper[a]	401	1.17
Gold[a]	318	1.28
Invar[b]	13.9	0.037
Iron[d]	79.1	0.228
Armco Iron[a]	72.8	0.207
Carbon Steel[b]	60.5	0.145
Lead[a]	35.3	0.243
Mercury[a]	8.3	0.044
Monel metal[b]	22.7	0.065
Nichrome[b]	12.9	0.040
Nickel[a]	91.6	0.230
Platinum[a]	71.6	0.252
Silver[a]	429	1.74
Titanium (polycrystalline)[a]	21.9	0.093
Tungsten[a]		
298.2 K	174	0.664
1500 K	107	0.354
3000 K	91.4	0.263
Bonded silicon carbide[e]	40.0	0.020
Titanium carbide[e]	31.5	0.120
Diamond (type I)[a]	907	5.09
Graphite (ATJ, along molding pressure)[a]	98.2	0.80
Quartz glass[f]		
289.2 K	1.38	0.0083
500 K	1.62	0.0077
1000 K	2.87	0.0114
Limestone (Bedford, IN)[g]	2.20	0.010

Table 6.6 *Continued.*

Material	Conductivity κ W/m·K	Diffusivity D $10^{-4}\,\mathrm{m^2/s}$
Soapstone[e]	3.35	0.015
MgO [f]	48.7	0.137
Al_2O_3 (sapphire)[e]	46.4	0.149
Al_2O_3 (polycrystalline)[e]	36.3	0.118
Mica[e]	0.79	0.0031
Glass (Pyrex)[f]	1.10	0.0064
Air[h]	0.026	0.202
Asbestos (loose fiber)[e]	0.16	0.0023
Firebrick[e]	0.75	0.0033
Concrete (cement, sand, and gravel)[e]	1.82	0.0090
Cork[e]	0.042	0.0014
Paraffin[e]	0.454	0.00197
Water[i]		
saturated liquid	0.609	0.0014
saturated vapor	0.0186	4.08
Wood (Mahogany)		
across grain	0.161	0.0023
along grain	0.310	0.0044

[a] C. Y. Ho, R. W. Powell, and P. E. Lilly, Thermal Conductivity of the Elements: A Comprehensive Review [J. Phys. Chem. Ref. Data, **3**, Suppl. 1 (1974)]; Y. S. Touloukian, R. W. Powell, C. Y. Ho, and M. C. Nicolaou, *Thermal Diffusivity*, vol. 10 of Thermophysical Properties of Matter: The TPRC Data Series (IFI/Plenum, New York, 1973).

[b] Y. S. Touloukian, R. W. Powell, C. Y. Ho, and P. G. Clemens, *Thermal Conductivity: Metallic Elements and Alloys*, Thermophysical Properties of Matter: The TPRC Data Series Vol. 1 (IFI/Plenum, New York, 1970).

[c] C. Y. Ho, M. W. Ackerman, K. Y. Wu, S. G. Oh, and T. N. Havill, Thermal Conductivity of Ten Selected Binary Alloy Systems [J. Phys Chem. Ref. Data, **7**, 959 (1978)].

[d] Y. S. Touloukian *et al.*, *Properties of Selected Ferrous Alloying Elements*, CINDAS Data Series on Material Properties Vol. III-1 (McGraw-Hill, New York, 1981).

[e] Y. S. Touloukian, R. W. Powell, C. Y. Ho, and P. G. Clemens, *Thermal Conductivity: Nonmetallic Solids*, vol. 2 of Thermophysical Properties of Matter: The TPRC Data Series (IFI/Plenum, New York, 1970).

[f] R. W. Powell, C. Y. Ho, and P. E. Lilly, *Thermal Conductivity of Selected Materials*, National Standard Reference Data Series, National Bureau of Standards, NSRDS-NBS 8 (1966).

[g] P. D. Desai, R. A. Navarro, S. E. Hasan. C. Y. Ho, D. P. Dewitt, and T. R. West, *The Thermophysical Properties of Selected Rocks*, Purdue University, TPRS Report 23 (1974).

[h] Y. S. Touloukian, P. E. Lilly, S. G. Saxena, *Thermal Conductivity: Nonmetallic Liquids and Gases*, Thermophysical Properties of Matter: The TPRC Data Series Vol. 3 (IFI/Plenum, New York, 1970).

[i] P. E. Lilly, T. Makita, and Y. Tanaka, *Properties of Inorganic and Organic Fluids*, CINDAS Data Series on Material Properties Vol. V-1 (Hemisphere, New York, 1988).

6.5.2   Radiative heat transfer

The energy radiated per unit area between wavelengths λ and $\lambda + \Delta\lambda$, integrated over the hemisphere (2π steradians) is

$$\Delta\Phi(\lambda) = \pi\epsilon_\lambda J_\lambda \Delta\lambda,$$

where ϵ_λ is the emissivity of the surface (the ratio of the emission of the surface to that of a black body at the same temperature). Planck's law for the emission of a black body,

$$J_\lambda = \frac{c_1}{\lambda^5(e^{c_2/\lambda T} - 1)}$$

gives the distribution of energy of the heat spectrum:

$$c_1 = 3.741\,7749(22) \times 10^{-16}\,\text{W·m}^2, \qquad c_2 = 0.014\,387\,69(12)\,\text{m·K}$$

The total heat lost per unit area of a body with emissivity ϵ_T is given by Stefan's radiation formula

$$\Phi = \epsilon_T \sigma T^4, \qquad \sigma = 5.67051(19) \times 10^{-8}\,\text{W·m}^{-2}\text{·K}^{-4},$$

where ϵ_T is the emissivity averaged over all wavelengths. The heat emitted from a surface of area A into a cone defined by the solid angle $d\Omega$ at an angle θ from the normal to the surface is

$$d\Phi = \frac{A\cos\theta}{\pi}\epsilon_T \sigma T^4.$$

The heat transfer per unit area by radiation between two parallel surfaces at absolute temperatures T_1 and T_2 separated by a distance small compared to the dimensions of the surfaces is

$$W = \sigma(T_2^4 - T_1^4).$$

Emissivity: The emissivity ϵ_T is very nearly unity for porous nonmetallic substances. For nonmetallic pigment paints, $\epsilon_T \approx 1$, but for aluminum paint near room temperature ϵ_T varies between 0.3 and 0.5. For metals the emissivity is temperature dependent; for clean metallic surfaces this can be approximated by a power law, expressed in the form

$$\epsilon_T = \epsilon_T(T_o)\left(\frac{T}{T_o}\right)^m$$

or, for the total emission,

$$\Phi = M\left(\frac{T}{T_o}\right)^{4+m}.$$

Table 6.7 Radiation constants of metals. $T_{\circ} = 1000\,\mathrm{K}$.
Data adapted from Ref. 5.

Metal	Temperature Range	m	$\epsilon_T(1000\ \mathrm{K})$	$M/(\mathrm{W \cdot m^2})$
Silver	610 − 980	0.1	0.11	6 000
Platinum	640 − 1150	1.0	0.41	23 000
Nickel	463 − 1280	0.65	0.16	8 900
Iron	700 − 1300	1.55	0.25	14 300
Nichrome	325 − 1310	0.1	0.63	36 000

References

1   N. W. Ashcroft and N. D. Mermin, *Solid State Physics*, 4th ed. (Wiley, New York, 1971).

2   *Handbook of Physics*, 2nd ed., edited by E. U. Condon and Hugh Odishaw (McGraw-Hill, New York, 1967).

3   *American Institute of Physics Handbook*, 3rd ed., edited by Dwight E. Gray (McGraw-Hill, New York, 1976).

4   *Mark's Mechanical and Engineering Handbook*, 7th ed., edited by T. Baumeister (McGraw-Hill, New York, 1967).

5   J. Strong, *Procedures in Experimental Physics* (Prentice-Hall, Englewood Cliffs, NJ, 1938).

6   B. N. Taylor, *Guide for the Use of the International System of Units, The Modernized Metric System*, NIST SP 811 (1995 edition), U.S. Department of Commerce, National Institute of Standards and Technology, Gaithersburg, MD 20899.

7   B. Goldfarb and F. R. Follett, *Units for Magnetic Properties*, NBS SP 696, U.S. Department of Commerce, National Institute of Standards and Technology, Boulder, CO 80303.

8   H. Preston-Thomas, The International Temperature Scale of 1990 (ITS-90) [Metrologia, **27**, 3 (1990)] [Official French text: BIPM Proc.-Verb. Com. Int. Poids et Mesures, 1989, **57**, T1-21].

7

PROPERTIES OF MATTER

7.1 Elementary particles

Table 7.1 Elementary particles. For antiparticles the signs of all quantum numbers and of the charge are reversed.

Charge		Spin $J = \frac{1}{2}$, Parity $= +$		
		LEPTONS, lepton number $= 1$		
		electron	muon	tauon
		e	μ	τ
$-e$	mc^2	0.510 999 MeV	105.658 MeV	1777 MeV
			neutrinos	
		ν_e	ν_μ	ν_τ
0	mc^2	$< 7\,\mathrm{eV}$	$< 270\,\mathrm{KeV}$	$< 31\,\mathrm{MeV}$
		There is no conclusive evidence that neutrinos have nonzero masses.		
		QUARKS, baryon number $= \frac{1}{3}$		
		u	c	t
$+\frac{2}{3}e$	mc^2	2–8 MeV	1.0–1.6 MeV	$\approx 182\,\mathrm{GeV}$
		$I_z = +\frac{1}{2}$	$C = +1$	$T = +1$
		d	s	b
$-\frac{1}{3}e$	mc^2	5–18 MeV	100–300 MeV	4.1–4.5 GeV
		$I_z = -\frac{1}{2}$	$S = -1$	$B = -1$

Table 7.1 *Continued.*

		FORCES		
Type	Relative Strength	Carrier	Charge	Mass
Strong	1	gluon, g	0	0
		π^0	0	134.976 MeV
		$\pi^\pm$	$\pm e$	139.57 MeV
Electro-Magnetic	$\alpha \approx 1/137$	γ	0	$0 \ (< 3{\times}10^{-27}\,\mathrm{eV})$
Weak	$\approx 10^{-14}$	$W^\pm$	$\pm e$	80.41 GeV
		Z^0	0	91.187 GeV
Gravitation	$\approx 10^{-38}$	graviton	0	0

7.2   The elements

7.2.1   Atomic weights

A chemical element is, in general, a combination of isotopes. The atomic weight $A_r(E)$ of element E is the mean relative atomic mass of the nuclides ^{A_k}E of given atomic number Z that constitute the element,

$$A_r(E) = \sum_k f(^{A_k}E) M_r(^{A_k}E), \qquad M_r(^{A_k}E) = \frac{m(^{A_k}E)}{m_u},$$

where Z and A are the atomic number and mass number of a nuclide (neutral atom in its electronic and nuclear ground state) and $f(^{A_k}E)$ is the isotopic abundance of the nuclide; m_u is the atomic mass unit [$m_u = \frac{1}{12}m(^{12}C)$].

Atomic weights are not fixed physical constants because (particularly in the lighter elements) the isotopic composition can vary depending on the source of the material and the treatment to which it is subjected. The primary uncertainty in the standard value of the atomic weight of an element (unless it is mononuclidic) is the isotopic abundances since the relative masses of stable nuclides are determined from mass spectroscopy and reaction energies to better than parts in 10^7.

In the following table, the digit in parentheses indicates the uncertainty of the last significant digit to which it is attributed. It serves a double purpose; for most elements it represents the uncertainty of the analytic determination, corresponding to an estimated two or three standard deviations, but for those entries with the notes g, m, or r, it reflects a real variability of the isotopic abundances of the elements as they may be found in the laboratory:

g Geological specimens with abnormal composition are known; the atomic weight of such a specimen may differ from the standard atomic weight by more than the indicated uncertainty.

m The isotopic composition can be modified by fractionation during processing and purification. Substantial deviations of the atomic weight from the standard value can occur.

r The range of isotopic composition of normal terrestrial material limits the precision of the given value. The stated uncertainty represents the variability of the atomic weight in readily available sources.

Table 7.2 Atomic weights of the elements, 1993. *Standard atomic weights* are for the isotopic composition of normally available terrestrial sources, but may not apply to exotic samples. Following each element name are the atomic numbers of stable isotopes of that element. If a more precise atomic weight is required, an isotopic analysis of the specific specimen is required.

Z	Name	Symbol	Atomic weight	Notes
1	Hydrogen (1,2)	H	1.007 94(7)	g,m,r
2	Helium (3,4)	He	4.0026 02(2)	g, r
3	Lithium (6,7)	Li	6.941(2)[a]	g,m,r
4	Beryllium (9)	Be	9.012 182(3)	
5	Boron (10,11)	B	10.811(5)	g,m,r
6	Carbon (12,13)	C	12.011(1)	g, r
7	Nitrogen (14,15)	N	14.006 74(7)	g, r
8	Oxygen (16,17,18)	O	15.999 4(3)	g, r
9	Fluorine (19)	Fl	18.998 4032(9)	
10	Neon (20,21,22)	Ne	20.179 7(6)	g,m
11	Sodium (23)	Na	22.989 76(6)	
12	Magnesium (24,25,26)	Mg	24.305 0(6)	
13	Aluminum (*Aluminium*) (27)	Al	26.981 539(5)	
14	Silicon (28,29,30)	Si	28.085 5(3)	m
15	Phosphorus (31)	P	30.973 762(4)	
16	Sulfur (32,33,34,36)	S	32.066(6)	g,m
17	Chlorine (35,37)	Cl	35.452 7(9)	m
18	Argon (39,38,40)	Ar	39.948(1)	g, r
19	Potassium (39,40[b],41)	K	39.098 3(1)	g
20	Calcium (40,42,43,44,46)	Ca	40.078(4)	g
21	Scandium (45)	Sc	44.955 910(9)	
22	Titanium (46,47,48,49,50)	Ti	47.867(1)	
23	Vanadium (50,51)	V	50.941 5(1)	
24	Chromium (50,52,53,54)	Cr	51.996 1(6)	
25	Manganese (52)	Mn	54.938 05(1)	

Table 7.2 *Continued.*

Z	Name	Symbol	Atomic weight	Notes
26	Iron (54,56,57,58)	Fe	55.845(2)	
27	Cobalt (59)	Co	58.933 20(1)	
28	Nickel (58,60,61,62,64)	Ni	58.693 4(2)	
29	Copper (63,65)	Cu	63.546(3)	r
30	Zinc (64,66,67,68,70)	Zn	65.39(2)	
31	Gallium (69,71)	Ga	69.723(1)	
32	Germanium (70,72,73,74,76)	Ge	72.61(2)	
33	Arsenic (75)	As	74.921 59(2)	
34	Selenium (74,76,77,78,80,82)	Se	78.96(3)	
35	Bromine (79,81)	Br	79.904(1)	
36	Krypton (78,80,82,83,84)	Kr	83.80(1)	g
37	Rubidium (85,87)	Ru	85.467 8(3)	g
38	Strontium (84,86,87,88)	Sr	87.62(1)	g, r
39	Yttrium (89)	Y	88.905 85(2)	
40	Zirconium (90,91,92,94,96)	Ni	91.224(2)	g
41	Niobium (93)	Nb	92.906 38(2)	
42	Molybdenum (92,94,95,96,98,100)	Mo	95.94(1)	g
43	Technetium	Tc	c	
44	Ruthenium (96,98,99,100,101,102,104)	Ru	101.07(2)	g
45	Rhodium (103)	Rh	102.905 50(3)	
46	Palladium (102,104,105,106,108,110)	Pa	106.42(1)	g
47	Silver (107,109)	Ag	107.868 2(2)	g
48	Cadmium (106,108,110,111,112,113,114)	Cd	112.411(8)	g
49	Indium (113,115)	In	114.818(3)	
50	Tin (112,114,115,116,117,118,119,120,122,124)			
		Sn	118.710(7)	g
51	Antimony (121,123)	Sb	121.760(1)	g
52	Tellurium (120,122,123,124,125,126,128,130)			
		Te	127.60(3)	g
53	Iodine (127)	I	126.904 47(3)	
54	Xenon (124,126,128,129,130,132,134,136)	Xe	131.29(2)	g,m
55	Cesium (*Caesium*) (133)	Cs	132.905 43(5)	
56	Barium (130,132,134,135,136)	Ba	137.327(7)	
57	Lanthanum (138,139)	La	138.905 5(2)	g
58	Cerium (136,138,140,142)	Ce	140.115(2)	g
59	Praesodymium (141)	Pr	140.907 65(3)	
60	Neodymium (143,144,145,146,148,150)	Nd	144.24(3)	g
61	Promethium	Pm	c	
62	Samarium (144,147^d,148^d,149^d,150)	Sa	150.36(3)	g
63	Europium (151,153)	Eu	151.965(9)	g
64	Gadolinium (152,154,155,156,157,158,160)			
		Gd	157.25(3)	g
65	Terbium (159)	Tb	158.925 34(3)	
66	Dysprosium (156,158,160,161,162,163,164)			
		Dy	162.50(3)	g
67	Holmium (165)	Ho	164.930 32(3)	
68	Erbium (162,164,166,167,168,170)	Er	167.26(3)	g

Table 7.2 *Continued.*

Z	Name	Symbol	Atomic weight	Notes
69	Thulium (169)	Tm	168.934 21(3)	
70	Ytterbium (168,170,171,172,173,174,176)	Yb	173.04(3)	g
71	Lutetium (175,176)	Lu	174.967(1)	g
72	Hafnium (174,176,177,178,179,180)	Hf	178.49(2)	
73	Tantalum (180,181)	Ta	180.947 9(1)	
74	Tungsten (180,182,183,184,186)	W	183.84(1)	
75	Rhenium (185,187)	Re	186.2407(1)	
76	Osmium (184,186,187,188,190,192)	Os	190.23(3)	g
77	Iridium (191,193)	Ir	192.217(3)	
78	Platinum (190,192,194,195,196,198)	Pt	195.08(3)	
79	Gold (197)	Au	196.966 54(3)	
80	Mercury (196,198,199,200,201,202,204)	Hg	200.59(2)	
81	Thallium (203,205)	Th	204.383 3(2)	
82	Lead (204,206,207,208)	Pb	207.2(1)	g, r
83	Bismuth (209)	Bi	208.980 37(3)	
84	Polonium	Po	c	
85	Astatine	At	c	
86	Radon	Rn	c	
87	Francium	Fr	c	
88	Radium	Ra	c	
89	Actinium	Ac	c	
90	Thorium (232)	Th	232.038 1(1)^e	g
91	Protactinium (231)	Pa	231.035 88(2)^f	
92	Uranium (235,238)	U	238.028 9(1)^g	g,m
93	Neptunium	Np	c	
94	Plutonium	Pu	c	
95	Americium	Am	c	
96	Curium	Cm	c	
97	Berkelium	Bk	c	
98	Californium	Cf	c	
99	Einsteinium	Es	c	
100	Fermium	Fm	c	

[a] The atomic weight of Li commonly available in the laboratory can vary significantly (6.94–6.99) from this standard value.

[b] Isotope has a half-life $t_{1/2} = 1.25 \times 10^9$ yr.

[c] There are no stable isotopes; a chemical atomic weight for the element cannot be given.

[d] Isotopes have half-lives: 147, 1.08×10^{11} yr; 148, 7×10^{15} yr; 149, 10^{16} yr.

[e] Although Th has no stable nuclide the lifetime of ^{232}Th ($t_{1/2} = 1.4 \times 10^{10}$ yr) is sufficiently long that a present-day atomic weight can be given.

[f] ^{231}Pa ($t_{1/2} = 3.27 \times 10^4$ yr) is in equilibrium with the α, β decay of ^{235}U.

[g] Although U has no stable nuclide the lifetimes are sufficiently long that a present-day atomic weight can be given.

Table 7.3 Transfermium elements. An international consensus on names and symbols for elements 104-108 has not been achieved. Those marked (A) have been used by the Chemical Abstracting Board, those marked (C) have been proposed by the IUPAC Commission on Nomenclature in Inorganic Chemistry[a]. It is expected that a final agreement on nomenclature will be announced in August, 1997.

101 Mendelevium	Md	105 (A) Hahnium	Ha	107 (A) Nielsbohrium	Ns
102 Nobelium	No	(C) Joliotium	Jl	(C) Bohrium	Bh
103 Lawrencium	Lr	106 (A) Seaborgium	Sg	108 (A) Hassium	Hs
104 (A) Rutherfordium	Rf	(C) Rutherfordium	Rf	(C) Hahnium	Ha
(C) Dubnium	Db			109 Meitnerium	Mt

[a] *Pure & Appl. Chem.*, **66**(12), 2419–2421 (1994).

Table 7.4 The elements: electronic configuration, ionization potential, melting and boiling points, and density. The melting and boiling points are corrected to ITS-90.

Element	Electron configuration	Ground state	Ionization potential I/eV	Melting point (°C)	Boiling point (°C)	Density[a] ρ (kg/m^3)
1 H	1s	$^2S_{1/2}$	13.598	−259.15	−252.88	0.0838
2 He	1s^2	$^1S_0$	24.588	−272.2 [b]	−268.928	0.1662
3 Li	[He]2s	$^2S_{1/2}$	5.392	180.58	1342	534
4 Be	[He]2s^2	$^1S_0$	9.322	1278	2970[c]	1848
5 B	[Be]2p	$^2P^o_{1/2}$	8.298	2078	2350[d]	2340
6 C	[Be]2p^2	$^3P_0$	11.260	3650[d]		3516[e]
7 N	[Be]2p^3	$^4S^o_{4/2}$	14.534	−209.86	−195.8	1.1655
8 O	[Be]2p^4	$^3P_2$	13.618	−218.4	−182.954	1.3316
9 Fl	[Be]2p^5	$^2P^o_{3/2}$	17.422	−219.62	−188.13	1.580
10 Ne	[Be]2p^6	$^1S_0$	21.564	−248.67	−246.044	0.8414
11 Na	[Ne]3s	$^2S_{1/2}$	5.139	97.79	883.0	971
12 Mg	[Ne]3s^2	$^1S_0$	7.646	648.8	1090	1738
13 Al	[Mg]3p	$^2P^o_{1/2}$	5.986	660.323	2465	2702
14 Si	[Mg]3p^2	$^3P_0$	8.151	1410	2350	2330[e]
15 P	[Mg]3p^3	$^4S^o_{3/2}$	10.486	44.1[f]	280	1820[f]
16 S	[Mg]3p^4	$^3P_2$	10.360	112.8	444.661	1957
17 Cl	[Mg]3p^5	$^2P^o_{3/2}$	12.967	−101.01	−34.6	2.9738
18 Ar	[Mg]3p^6	$^1S_0$	15.759	−189.2	−185.7	1.662
19 K	[Ar]4s	$^2S_{1/2}$	4.341	63.23	760.3	862
20 Ca	[Ar]4s^2	$^1S_0$	6.113	839	1484	1550
21 Sc	[Ca]3d	$^2D_{3/2}$	6.562	1541	2830	2989[e]
22 Ti	[Ca]3d^2	$^3F_2$	6.820	1660	3285	4540
23 V	[Ca]3d^3	$^4F_{3/2}$	6.740	1890	3380	6110
24 Cr	[Ar]3d^{5}4s	$^7S_3$	6.766	1857	2670	7190
25 Mn	[Ca]3d^5	$^6S_{5/2}$	7.437	1244	1960	7200

Table 7.4 *Continued.*

Element	Electron configuration	Ground state	Ionization potential I/eV	Melting point (°C)	Boiling point (°C)	Density[a] ρ (kg/m^3)
26 Fe	[Ca]3d^6	^{5}D$_4$	7.870	1535	2750	7874
27 Co	[Ca]3d^7	^{4}F$_{9/2}$	7.864	1495	2870	8900
28 Ni	[Ca]3d^8	^{3}F$_4$	7.638	1453	2730	8902
29 Cu	[Ar]3d^{10}4s	^{2}S$_{1/2}$	7.478	1084.62	2565	8960
30 Zn	[Ca]3d^{10}	^{1}S$_0$	9.394	419.527	907	7133^e
31 Ga	[Zn]4p	^{2}P$^o_{3/2}$	5.999	29.7646	2402	5930
32 Ge	[Zn]4p^2	^{3}P$_0$	7.899	937.3	2828	5323
33 As	[Zn]4p^3	^{4}S$^o_{3/2}$	9.81	817^g	613^d	5730
34 Se	[Zn]4p^4	^{3}P$_2$	9.752	217	685.0	4820
35 Br	[Zn]4p^5	^{2}P$^o_{3/2}$	11.814	−7.2	58.76	3120
36 Kr	[Zn]4p^6	^{1}S$_0$	13.999	−156.6	−152.29	3.492
37 Rb	[Kr]5s	^{2}S$_{1/2}$	4.177	38.88	686	1532
38 Sr	[Kr]5s^2	^{1}S$_0$	5.695	769	1384	2540
39 Y	[Sr]4d	^{2}D$_{3/2}$	6.22	1522	3336	4469^e
40 Zr	[Sr]4d^2	^{3}F$_2$	6.84	1850	4375	6506
41 Nb	[Kr]4d^{4}5s	^{6}D$_{1/2}$	6.88	2468	4740	8570
42 Mo	[Kr]4d^{5}5s	^{7}S$_3$	7.099	2615	4610	10220
43 Tc	[Sr]4d^5	^{6}S$_{5/2}$	7.28	2173	4875	11500
44 Ru	[Kr]4d^{7}5s	^{5}F$_5$	7.37	2310	3900	12410
45 Rh	[Kr]4d^{8}5s	^{4}F$_{9/2}$	7.46	1965	3730	12410
46 Pd	[Kr]4d^{10}	^{1}S$_0$	8.34	1554	3140	12020
47 Ag	[Pd]5s	^{2}S$_{1/2}$	7.576	961.78	2210	10500
48 Cd	[Pd]5s^2	^{1}S$_0$	8.993	320.9	765	8650
49 In	[Cd]5p	^{2}P$^o_{1/2}$	5.786	156.9585	2080	7310
50 Sn	[Cd]5p^2	^{3}P$_0$	7.344	231.928	2270	7310
51 Sb	[Cd]5p^3	^{4}S$^o_{3/2}$	8.641	630.60	1750	6691
52 Te	[Cd]5p^4	^{3}P$_2$	9.009	449.4	990	6240
53 I	[Cd]5p^5	^{2}P$^o_{3/2}$	10.451	113.5	184.39	4930
54 Xe	[Cd]5p^6	^{1}S$_0$	12.130	−119.9	−107.1	5.486
55 Cs	[Xe]6s	^{2}S$_{1/2}$	3.894	28.39	669.4	1873
56 Ba	[Xe]6s^2	^{1}S$_0$	5.212	725	1640	3500
57 La	[Ba]5d	^{2}D$_{3/2}$	5.577	921	3455	6145^e
58 Ce	[Ba]4f5d	^{1}G$^o_{4/2}$	5.466	799	3425	6770^{e,i}
59 Pr	[Ba]4f^3	^{4}I$^o_{9/2}$	5.422	931	3510	6773^e
60 Nd	[Ba]4f^4	^{5}I$_4$	5.489	1021	3067	7008^e
61 Pm	[Ba]4f^5	^{6}H$^o_{5/2}$	5.554	1168	2700(?)	7220^e
62 Sa	[Ba]4f^6	^{7}F$_0$	5.631	1077	1790	7520^e
63 Eu	[Ba]4f^7	^{8}S$^o_{7/2}$	5.666	822	1597	5243^e
64 Gd	[Ba]4f^{7}5d	^{9}D$^o_2$	6.141	1313	3266	7901^e
65 Tb	[Ba]4f^9	^{6}H$^o_{15/2}$	5.852	1356	3123	8229
66 Dy	[Ba]4f^{10}	^{5}I$_8$	5.927	1409	2335	8550^e
67 Ho	[Ba]4f^{11}	^{4}I$^o_{15/2}$	6.018	1470	2720	8795^e
68 Er	[Ba]4f^{12}	^{3}H$_6$	6.101	1522	2510	9066^e
69 Tm	[Ba]4f^{13}	^{2}F$^o_{7/2}$	6.184	1545	1727	9321^e

Table 7.4 *Continued.*

Element	Electron configuration	Ground state	Ionization potential I/eV	Melting point (°C)	Boiling point (°C)	Density[a] ρ (kg/m^3)
70 Yb	[Ba]4f^{14}	$^1S_0$	6.254	820	1193	6965
71 Lu	[Yb]5d	$^2D_{3/2}$	5.426	1663	3315	9841[e]
72 Hf	[Yb]5d^2	$^3F_2$	6.65	2227	4620	13310
73 Ta	[Yb]5d^3	$^4F_{3/2}$	7.89	2995	5420	16654
74 W	[Yb]5d^4	$^5D_0$	7.98	3410	5660	19350
75 Re	[Yb]5d^5	$^6S_{5/2}$	7.88	3180	~5600	21020
76 Os	[Yb]5d^6	$^5D_4$	8.7	3045	5027	22570
77 Ir	[Yb]5d^7	$^4F_{9/2}$	9.1	2410	4130	22420
78 Pt	[Xe]4f^{14}5d^{9}6s	$^3D_3$	9.0	1772	3830	21450
79 Au	[Xe]4f^{14}5d^{10}6s	$^2S_{1/2}$	9.225	1064.18	3080	19300
80 Hg	[Yb]5d^{10}	$^1S_0$	10.437	−38.86	356.54	13546
81 Tl	[Hg]6p	$^2P^o_{1/2}$	6.108	303.5	1457	11850
82 Pb	[Hg]6p^2	$^3P_0$	7.416	327.462	1740	11350
83 Bi	[Hg]6p^3	$^4S^o_{3/2}$	7.289	271.3	1560	9747
84 Po	[Hg]6p^4	$^3P_2$	8.42	254	962	9320
85 At	[Hg]6p^5	$^2P^o_{3/2}$	8.8	302	337	
86 Rn	[Hg]6p^6	$^1S_0$	10.748	−71	−61.8	9.23
87 Fr	[Rn]7s	$^2S_{1/2}$	3.8			
88 Ra	[Rn]7s^2	$^1S_0$	5.279	700	1140	
89 Ac	[Ra]6d	$^2D_{3/2}$	5.17	1050	3200	10070
90 Th	[Ra]6d^2	$^3F_2$	6.08	1750	~4790	11720
91 Pa	[Rn]5f^{2}6d7s^2	$(4,\frac{3}{2})_{11/2}$	5.89	¡1600		15370
92 U	[Rn]5f^{3}6d7s^2	$(\frac{9}{2}\frac{3}{2})_6$	6.05	1132.0	3815	19050
93 Np	[Ra]5f^{4}6d7s^2	$(4,\frac{3}{2})_{11/2}$	6.19	640	3900	20250
94 Pu	[Ra]5f^6	$^7F_0$	6.06	641	3230	
95 Am	[Ra]5f^7	$^8S^o_{1/2}$	5.993	994	2606	
96 Cm	[Rn]5f^{7}6d7s^2	$(\frac{7}{2},\frac{3}{2})^o_2$	6.02	1340		
97 Bk	[Ra]5f^9	$^6H^o_{15/2}$	6.23			
98 Cf	[Ra]5f^{10}	$^5I_8$	6.30			
99 Es	[Ra]5f^{11}	$^4I^o_{15/2}$	6.42			
100 Fm	[Ra]5f^{12}	$^3H_6$	6.50			
101 Md	[Ra]5f^{13}	$^2F^o_{1/2}$	6.58			
102 No	[Ra]5f^{14}	$^1S_0$	6.65			
103 Lr	[Ra]5f^{14}6d	$^2D_{3/2}$	8.6			

[a] Densities at 20 °C and 1 atm = 101.325 kPa.

[b] At 2.6 MPa.

[c] At 670 Pa.

[d] Diamond; density of crystalline graphite, 2250 kg/m^3; amorphous graphite, 1600–1650 kg/m^3.

[e] At 25 °C.

[f] White phosphorus. Density of red phosphorus, 2200 kg/m^3; black phosphorus, 2250–2690 kg/m^3.

[g] At 2.8 MPa.

7.3   Densities

Table 7.5 Selected solids, liquids, and gases.

Substance	Density $\rho/(\text{kg/m}^3)$	Substance	Density $\rho/(\text{kg/m}^3)$
		Solids	
Carbon		Cork	220– 260
bulk graphite	1650–1800	Ebonite	1150
cryst. graphite	2250	Glass	2400–2800
Diamond	3514	Ice (0 °C)	917
Iron		Mica	2600–3200
pure	7880	Fused silica	2100–2200
wrought	7850	Paraffin	870–910
cast	7600	Wood (oven dry)	
Brass	8400–8700	ash	520–640
Bronze	8800–8900	balsa	120–200
Phosphor Bronze	8800	beech	650–670
Constantin	8880	elm	550–670
Invar	8000	lignum vitae	1100–1250
Manganin	8500	mahogany	540–670
Steel	7880	oak	670–980
Wood's metal[b]	9500–10500	teak, Indian	580–800
		Liquids	
Acetone	0.792		
Aniline	1.02	Aqueous solutions (1 mol/L)[a]	
Benzene	0.899		
Ether	0.736	$\frac{1}{2}H_2SO_4$	1030.4
Ethyl alcohol	0.791	HCl	1016.2
Glycerine	1.26	HNO$_3$	1032.2
Lubricating oil	0.90–0.92	NaOH	1041.4
Methyl alcohol	0.810	NaCl	1038.8
Olive oil	0.92	KOH	1048
Paraffin oil	0.8	KCl	1044.6
Turpentine	0.87		

Gases, $T = 273.15$ K, $(t = 0\,°C)$; $p = 101.325$ kPa (for densities of elemental gases, see Table **7.2**)			
Air [c]	1.2928	Hydrochloric acid (HCl)	1.639
Ammonia (NH$_3$)	0.7710	Hydrogen sulfide (H$_2$S)	1.539
Carbon monoxide (CO)	1.258	Methane (CH$_4$)	0.5547
Carbon dioxide (CO$_2$)	1.977		

[a]   Densities at 20 °C.

[b]   4Bi:2Pb:Sn:Cd.

[c]   Composition: 78.084 N$_2$, 20.946 O$_2$, 0.934 Ar, 0.033 CO$_2$, 0.002 Ne. Mean atomic weight: 28.965.

Table 7.6 Density of water and mercury.   In this table the density of water has been normalized to $999.9720\,\mathrm{kg/m^3}$ at $t = 3.98\ °C$. The absolute values have an uncertainty of at least $1\,\mathrm{g/m^3}$. The density of mercury is standardized for barometric use to $13\,595.1\,\mathrm{kg/m3}$ at $0\,°C$.

t °C	Water kg/m^3	t °C	Water kg/m^3	Mercury kg/m^3	t °C	Mercury kg/m^3
0	999.8395	0	999.840	13 595.1	120	13 304.0
1	999.8985	10	999.700	13 570.4	140	13 256.3
2	999.9399	20	999.203	13 545.8	160	13 208.9
3	999.9642	30	998.646	13 521.3	180	13 161.6
4	999.9720	40	992.21	13 497.0	200	13 114.4
5	999.9638	50	988.04	13 472.5	220	13 067.4
6	999.9402	60	983.21	13 448.2	240	13 020.8
7	999.9015	70	977.79	13 424.0	260	12 973.7
8	999.8482	80	971.80	13 400.0	280	12 926.9
9	999.7808	90	965.31	13 375.8	300	12 880.2
		100	958.35	13 351.8	350	13 762.4

7.3.1   Relative density, specific gravity

The maximum density of pure air-free water is

$$\rho_{\max} = 999.972\,\mathrm{kg/m^3}$$

at $T = 277.13\,\mathrm{K}$, $(3.98\,°C)$. For water and other liquids with densities near $1000\ \mathrm{kg/m^3}$, the specific gravity is

$$d = \rho/\rho(\mathrm{H_2O})_{\max}.$$

7.4   Viscosity

Couette flow: When a fluid is contained between two parallel plates separated by a distance δz, the force per unit area required to maintain a velocity difference δv between the plates is

$$\frac{dF}{dA} = \eta\frac{\Delta v}{\Delta z},$$

where η is the *dynamic viscosity*.

For an ideal gas of hard spheres, diameter d and mass m, the viscosity is

$$\eta = \frac{5}{16d^2}\sqrt{\frac{mkT}{\pi}},$$

which is independent of the pressure of the gas.

Poiseuille Flow: The pressure gradient required to maintain a flow Φ (volume flow rate) or Ψ (mass flow rate) of fluid in a circular pipe of internal radius a, (neglecting the boundary effects at the wall of the pipe) is

$$\frac{dp}{dx} = \eta \frac{8\Phi}{\pi a^4} = \nu \frac{8\Psi}{\pi a^4},$$

where $\nu = \eta/\rho$ is the kinematic viscosity.

Table 7.7 Viscosity of water and air.   ($1\,\text{Pa·s} = 10\,\text{poise}$)

	Water					Air		
t	η	ν	t	η	ν	t	η	ν
°C	mPa·s	mm²/s	°C	μPa·s	mm²/s	°C	μPa·s	mm²/s
0	1.792	1.792	0	17.09	13.2	220	26.58	37.1
10	1.308	1.308	20	18.08	15.0	240	27.33	39.7
20	1.005	1.007	40	19.04	16.9	260	28.06	42.4
30	0.801	0.804	60	19.97	18.8	280	28.77	45.1
40	0.656	0.611	80	20.88	20.9	300	29.46	48.1
50	0.549	0.556	100	21.75	33.0	320	30.14	50.7
60	0.469	0.477	120	2.260	25.2	240	30.80	53.5
70	0.406	0.415	140	23.44	27.4	360	31.46	56.5
80	0.357	0.367	160	24.25	29.8	380	32.12	59.5
90	0.317	0.328	180	25.05	32.2	400	32.77	62.5
100	0.284	0.296	200	25.82	34.6	420	33.40	65.6

The viscosity of water may be represented empirically (to better than 0.2 percent) by the following expression (adapted from Ref. 3)

$$\ln \frac{\eta(t)}{\eta(20\,°\text{C})} = -a(t - 20)\left[1 + \frac{b(t - 20)}{t + 98.5}\right], \qquad 0\,°\text{C} < t < 100\,°\text{C},$$

where $a = 24.475 \times 10^{-3}$, $b = -0.8787$.

Table 7.8 Viscosity of glycerol/water mixtures.　For a more complete listing of values, see Ref. 3.

Temp. $t/\,^{\circ}\mathrm{C}$	Glycerol fraction, by weight				
	20%	40%	60%	80%	100%
	mPa·s	mPa·s	mPa·s	mPa·s	mPa·s
0	3.44	8.25	29.9	255	12070
10	2.41	5.37	17.4	116	3900
20	1.76	3.72	10.8	60.1	1412
30	1.35	2.72	7.19	33.9	612
40	1.07	2.07	5.08	20.8	284
50	0.879	1.68	3.76	13.6	142
60	0.731	1.30	2.85	9.42	81.3
70	0.635	1.09	2.29	6.94	50.6
80		0.918	1.84	5.13	31.9
90		0.763	1.52	4.03	21.3
100		0.666	1.28	3.18	14.8

Table 7.9 Viscosity of mercury.

t $^{\circ}\mathrm{C}$	η μPa·s	ν $\mathrm{mm^2/s}$	t $^{\circ}\mathrm{C}$	η μPa·s	ν $\mathrm{mm^2/s}$	t $^{\circ}\mathrm{C}$	η μPa·s	ν $\mathrm{mm^2/s}$
−20	1855	0.1359	40	1450	0.1074	100	1240	0.0929
−10	1764	0.1295	50	1407	0.1044	150	1130	0.0854
0	1685	0.1239	60	1367	0.1016	200	1052	0.0802
10	1615	0.1190	70	1331	0.0991	250	995	0.0766
20	1554	0.1147	80	1298	0.0969	300	950	0.0738
30	1499	0.1109	90	1268	0.0948	350	914	0.0716

Table 7.10 Viscosity of various gases and liquids.　Viscosity, unless otherwise noted, at 20 °C.

Substance	Viscosity (mPa·s)	Substance	Viscosity (μPa·s)
Liquids		*Gases*	
Acetone	0.333	Ammonia	10.8
Alcohol		Argon	21.0
Ethyl	1.192	Carbon monoxide	18.4
Methyl	0.591	Carbon dioxide	16.0
Benzene	0.649	Chlorine	14.7
Carbon disulfide	0.367	Helium	18.9
Castor oil	0.234	Hydrogen	9.5

Table 7.10 *Continued.*

Substance	Viscosity (mPa·s)	Substance	Viscosity (µPa·s)
Ether	0.830	Hydrogen sulfide	13.0
Linseed oil	986	Hydrogen chloride	14.0
Nitric acid	1.770[a]	Krypton	23.0
Olive oil	84	Methane	12.0
Sulfuric acid	22	Neon	29.8
Turpentine	1.49	Nitric oxide	18.6
Xylol		Nitrogen	18.4
Ortho	0.807	Oxygen	20.9
Meta	0.615	Sulfur dioxide	13.8
Para	0.643	Water vapor	9.8
		Xenon	21.1

[a] At 10 °C.

7.5   Surface tension

Table 7.11 Surface tension of various liquids.   Except as noted, $T = 20\,°C$. Surface tension is for an air/liquid interface.
$$(1\,\mathrm{mJ/m^2} = 10^{-3}\,\mathrm{N/m} = 1\,\mathrm{dyne/cm} = 1\,\mathrm{erg/cm^2})$$

Substance	Surface tension $\gamma/(\mathrm{mJ/m^2})$		Surface tension $\gamma/(\mathrm{mJ/m^2})$
		Water	
Acetone	23.7	Temp.	
Ethyl alcohol	22.3	(°C)	
Methyl alcohol	22.6	10	74.22
Amyl Acetate	24.7	20	72.75
Benzene	28.9	30	71.18
Chloroform	27.2	40	69.56
Copper (1404 °C)	1100	50	67.91
Lead (350 °C)	453	60	66.18
Mercury	472 [a]	70	64.4
Olive oil	33	80	62.6
Nitric acid	41	90	60.8
Sulfuric acid	55	100	58.9
Turpentine	27		

[a] In vacuum (Hg vapor); surface tension in air, $\gamma \approx 435\,\mathrm{mJ/m^2}$.

Table 7.12 Surface tension of aqueous solutions.   The quantity tabulated is the increase in surface tension associated with the addition of 1 mole per liter of solute to pure water.

Solute	$\dfrac{\Delta\gamma}{\mu N/m \big/ mol/L}$	Solute	$\dfrac{\Delta\gamma}{\mu N/m \big/ mol/L}$
CaCl$_2$	0.32	NaCl	0.16
CuSO$_4$	0.25	NaOH	0.20
KCl	0.14	NH$_3$Cl	0.15
KOH	0.18		

7.6   Vapor pressure

Table 7.13 Vapor pressure of water and mercury.

Temperature $t/(°C)$	Water kPa	Water mmHg	Mercury Pa	Mercury mmHg
−195 [a]	...	...	3×10^{-25}	2×10^{-27}
−78.5 [b]	8×10^{-5}	6×10^{-4}	4×10^{-7}	3×10^{-9}
−20	0.105 [c]	0.79 [c]	...	...
−10	0.263 [c]	1.97 [c]	...	...
0	0.610	4.58	0.053	0.0004
10	1.23	9.2	...	...
20	2.33	17.5	0.17	0.0013
30	4.24	31.8	...	...
40	7.37	55.3	0.8	0.006
50	12.33	92.5	...	...
60	19.92	149.4	4	0.03
70	31.2	234	...	...
80	47.3	355	12	0.09
90	70.1	526	...	...
100	101.3	760	37	0.28
150	477.3	3580	390	2.9
200	1560	11 700	2 400	18
300	...	...	32 900	247

[a] Liquid air.
[b] CO$_2$ sublimation (dry ice).
[c] Ice.

Table 7.14 Vapor pressure of selected substances.
Vapor pressure at 20 °C.

Substance	kPa	Substance	kPa
Acetic acid	1.55	Chloroform	21.5
Acetone	24.7	Ethanol	5.93
Benzene	9.95	Ethyl acetate	9.71
Bromine	22.9	Ethyl ether	58.7
Carbon disulfide	39.7	Hydrogen sulfide	1700
Carbon dioxide	5720	Methanol	11.83
Carbon tetrachloride	12.1	Sulfur dioxide	328

Table 7.15 Wet and dry bulb hygrometer table. (Ventilated-type hygrometer) The tabulated values are percent relative humidity.

Depression of wet bulb $\Delta t/$ °C	Dry-bulb temperature $t/$ °C								
	0	5	10	15	20	25	30	35	40
1	81	87	88	89	90	92	93	93	94
2	64	72	76	80	82	85	86	87	88
3	46	59	66	71	74	77	79	81	82
4	29	45	55	62	66	70	73	75	76
5	13	33	44	53	59	63	67	70	72
6		21	34	44	52	57	61	64	66
7		9	25	36	45	50	55	59	61
8			15	28	38	44	50	54	56
9			6	20	30	38	44	50	52
10				13	24	33	39	44	48

7.7 Compressibility

Table 7.16 Table of relative volume V/V_o and compressibility κ of selected liquids. $\kappa = (1/V)\partial V/\partial p$; V_o is the volume of the liquid at $t = 0$ °C and $p = 101.325$ kPa.

Pressure [a]	0.1 MPa		50 MPa		100 MPa	
Substance	V/V_o	$\dfrac{\kappa}{\text{GPa}^{-1}}$	V/V_o	$\dfrac{\kappa}{\text{GPa}^{-1}}$	V/V_o	$\dfrac{\kappa}{\text{GPa}^{-1}}$
Acetone	1.0279	1.19	0.9829	0.63	0.9553	0.54
Carbon disulfide	1.0235	0.91	0.9865	0.59	0.9586	0.50
Ethanol	1.0212	1.04	0.9794	0.64	0.9506	0.57

Table 7.16 *Continued.*

Pressure [a]	0.1 MPa		50 MPa		100 MPa	
Substance	V/V_o	$\dfrac{\kappa}{\mathrm{GPa}^{-1}}$	V/V_o	$\dfrac{\kappa}{\mathrm{GPa}^{-1}}$	V/V_o	$\dfrac{\kappa}{\mathrm{GPa}^{-1}}$
Ether	1.0315	1.82	0.9681	0.90	0.9363	0.65
Mercury [b]	1.00398	0.0401	1.00202	0.0396	1.00007	0.0391
Methanol	1.0283	1.12	0.9823	0.66	0.9530	0.56
Propanol	1.0173	0.91	0.9780	0.68	0.9498	0.49
Water	1.0016	0.461	0.9808	0.396	0.9630	0.356

[a]　$\mathrm{MPa} = 10^6\,\mathrm{N/m^2}$; $\mathrm{GPa} = 10^9\,\mathrm{N/m^2}$.
[b]　Volume and compressibility at $t = 22\,°\mathrm{C}$.

7.8　Materials for nuclear physics

Table 7.17 Gamma-ray shielding.　Linear attenuation coefficients for various materials.

Material	Density (kg/m^3)	Linear attenuation coefficient μ/m^{-1}		
		1 MeV	3 MeV	6 MeV
Air	1.293	0.00766	0.00430	0.00304
Aluminum	2700	16.6	9.53	7.18
Ammonia (liquid)	771	6.12	3.22	2.21
Beryllium	1850	10.4	5.79	3.92
Beryllium carbide	1900	11.2	6.27	4.29
Beryllium oxide (hot-pressed)	2300	14.0	7.89	5.52
Bismuth	9800	70.0	40.9	44.0
Boral	2530	15.3	8.65	6.78
Boron (amorphous)	2450	14.4	7.91	6.79
Boron carbide (hot pressed)	2500	15.0	8.25	6.75
Bricks				
fire clay	2050	12.9	7.38	5.43
kaolin	2100	13.2	7.50	5.52
silica	1780	11.3	6.46	4.73
Carbon	2250	14.3	8.01	5.54
Clay	2200	13.0	8.01	5.90
Cement				
(1 Portland cement : 3 sand)	2070	13.3	7.60	5.59
borated colemanite	1950	12.8	7.25	5.28
Concrete				
barytes	3500	21.3	12.7	11.0
barytes–boron frits	3250	19.9	11.9	10.1
barytes–limonite	3250	20.0	11.9	9.91
barytes–limonite–colemanite	3100	18.9	11.2	9.39

Table 7.17 *Continued.*

Material	Density (kg/m³)	Linear attenuation coefficient μ/m^{-1}		
		1 MeV	3 MeV	6 MeV
iron–Portland[a]	6000	36.4	21.5	18.1
MO (ORNL mixture)	2200	37.4	22.2	18.4
Portland concrete[b]	2200	14.1	8.05	5.92
Flesh[c]	1000	6.99	3.93	2.74
Fuel oil	890	7.12	3.50	2.39
Gasoline	739	5.37	2.99	2.03
Glass				
borosilicate	2230	14.1	8.05	5.91
lead (Hi-D)	6400	43.9	25.7	25.7
plate (average)	2400	15.2	8.62	6.29
Iron	7860	47.0	28.2	24.0
Lead	11340	79.7	46.8	50.5
Lithium hydride (pressed powder)	700	4.44	2.39	1.72
Lucite	1190	8.16	4.57	3.17
Paraffin	890	6.46	3.60	2.46
Rock				
granite	2450	15.5	8.86	6.54
limestone	2910	18.7	10.9	8.24
sandstone	2400	15.2	8.71	6.41
Rubber				
copolymer	915	6.62	3.70	2.54
natural	920	6.52	3.64	2.48
neoprene	1230	8.13	4.62	3.33
Sand	2200	14.0	8.25	5.87
Steel				
347 stainless	7800	46.2	27.9	23.6
1 % carbon	7830	46.0	27.6	23.4
Uranium	18700	146	81.3	88.1
Uranium hydride	11500	90.3	50.4	54.2
Water	1000	7.06	3.96	2.77
Wood				
ash	510	3.45	1.93	1.34
oak	770	5.21	2.93	2.03
white pine	670	4.52	2.53	1.75

[a] Elemental composition (% by weight): O 52.9; Si 33.7; Ca 4.4; Al 3.4; Na 1.6; Fe 1.4; K 1.3; H 1.0; Mg 0.2; C 0.1.

[b] Composition: 1 portland cement : 2 sand : 4 gravel.

[c] Composition (% by weight): O 65.99; C 18.27; H 10.15; N 3. Ca 1.52; P 1.02.

Table 7.18 Radiation damage dose for typical electrical insulating materials.

Material	Threshold damage dose D_{th}/MGy	25% damage dose D_{25}/MGy
Teflon	0.00017	0.00037
Formvar	0.16	0.82
Mylar	0.30	1.20
Mica	20	~150
Typical Epoxies (unfilled)	2	32
Epoxy, mineral filled with glass fiber reinforcing	80	500

References

[1] IUPAC Commission on Atomic Weights and Isotopic Abundances, *Atomic weights of the elements, 1993*, Pure and Applied Chemistry, **66**, 2423-2444 (1994).

[2] G. S. Kell, J. Chem. Eng. Data **20**(1), 97 (1975).

[3] R. C. Hardy and R. L. Cottington, Journ. Res. Natl. Bur. Stands. **42**, 573 (1949), and other unpublished results.

[4] R. J. Donnelly, *Fluid Dynamics* (section 12 of *A Physicist's Desk Reference*, H. L. Anderson, editor, American Institute of Physics, New York, 1989).

8

MATHEMATICAL FORMULAS

8.1 Algebraic equations

8.1.1 Quadratic equation

If $a \neq 0$, the two roots of $ax^2 + bx + c = 0$ are

$$x = \frac{-b \pm (b^2 - 4ac)^{1/2}}{2a} = \frac{-2c}{b \pm (b^2 - 4ac)^{1/2}},$$

$$x_1 = -\frac{bQ}{2a}, \qquad x_2 = -\frac{2c}{bQ}, \qquad Q = 1 + \sqrt{1 - \frac{4ac}{b^2}}.$$

There will be one double root if $D = b^2 - 4ac = 0$, $(Q = 1)$; if the coefficients are real, the two roots will be real if $D > 0$ and complex conjugates if $D < 0$.

The square root of the complex number $z = x + iy$ (x and y real) is

$$\sqrt{z} = \pm(a + ib)$$

with

$$a = \sqrt{\frac{1}{2}\left(\sqrt{x^2 + y^2} + x\right)}, \qquad b = \sqrt{\frac{1}{2}\left(\sqrt{x^2 + y^2} - x\right)}.$$

To ensure that the roots are in the correct quadrants of the complex plane and to avoid loss of numerical accuracy one should use these expressions to compute a if $x > 0$ or b if $x < 0$, and obtain the other quantity from the relation $2ab = y$.

8.1.2 Cubic equation

The cubic equation $ax^3 + bx^2 + cx + d = 0$ can be solved in closed algebraic form (Cardan, 1545) but the solution is computationally cumbersome, particularly for real roots; it can be more conveniently expressed in terms of trigonometric or hyperbolic functions:

$$\cos\phi = -B/A^{3/2}, \qquad x_k = 2\sqrt{A}\cos\left[\tfrac{1}{3}(\phi + 2\pi k)\right] - \frac{b}{3a}, \qquad k = 1, 2, 3$$

or

$$x_{1,2} = -\frac{b}{3a} - \sqrt{A}\left[\cos\tfrac{1}{3}\phi \pm \sqrt{3}\sin\tfrac{1}{3}\phi\right] \qquad x_3 = -\frac{b}{3a} + 2\sqrt{A}\cos\tfrac{1}{3}\phi$$

where

$$A = \left(\frac{b}{3a}\right)^2 - \frac{c}{3a}, \quad B = \left(\frac{b}{3a}\right)^3 - \frac{bc}{6a^2} + \frac{d}{2a}.$$

If the coefficients are real there is at least one real root; if $D = B^2 - A^3 \le 0$ there are three real roots; if $D = 0$, there is a double root (a triple root if $A = B = 0$); if $D > 0$ there is one real root and a complex conjugate pair.

8.1.3 Quartic equation

Given $ax^4 + bx^3 + cx^2 + dx + e = 0$.

Set $x = y - b/4a$ to obtain the reduced quartic

$$y^4 + Ay^2 + By + C = 0,$$

where

$$A = \frac{c}{a} - \frac{3b^2}{8a^2}, \quad B = \frac{d}{a} - \frac{cb}{2a^2} + \frac{b^3}{8a^3}, \quad C = \frac{e}{a} - \frac{bd}{4a^2} + \frac{cb^2}{16a^3} - \frac{3b^4}{256a^4}$$

and let the three roots of

$$z^3 + Az^2 + (A^2 - 4C)z - B^2 = 0$$

be u_1^2, u_2^2, and u_3^2. Then the four roots of the quartic are

$$x_1 = \frac{1}{2}(+u_1 + u_2 + u_3) - \frac{a}{4b}, \qquad x_3 = \frac{1}{2}(-u_1 + u_2 - u_3) - \frac{a}{4b},$$

$$x_2 = \frac{1}{2}(+u_1 - u_2 - u_3) - \frac{a}{4b}, \qquad x_4 = \frac{1}{2}(-u_1 - u_2 + u_3) - \frac{a}{4b}.$$

8.2   Factorials

Definition: $0! = 1,$ $n! = n \cdot (n-1)!,$

$1! = 1,$ $2! = 2,$ $3! = 6,$ $4! = 24,$ $5! = 120,$ $\ldots$, $10! = 3\,628\,800,$ $\ldots$,

$$(2n)!! = 2 \cdot 4 \cdot 6 \cdots (2n) = 2^n \cdot n!,$$

$$(2n+1)!! = 1 \cdot 3 \cdot 5 \cdots (2n+1) = \frac{(2n+1)!}{(2n)!!} = \frac{(2n+1)!}{2^n \cdot n!},$$

$$0!! \equiv 1, \qquad (-1)!! \equiv 1,$$

Stirling's Formula: $\displaystyle \lim_{n \to \infty} \frac{n! e^n}{\sqrt{n}\, n^n} = \sqrt{2\pi}\,.$

This gives approximate values of $n!$ for large n. An asymptotically better approximation that is also correct for $n = 0$ is given by

$$n! \approx \sqrt{2\pi n + 1}\, n^n e^{-n}.$$

For additional relations, see Sec. 8.14 Gamma function.

8.3   Binomial theorem

If n is a positive integer,

$$(1+z)^n = 1 + nz + \frac{n(n-1)z^2}{2!} + \frac{n(n-1)(n-2)z^3}{3!} + \cdots + \frac{n! z^r}{r!(n-r)!} + \cdots .$$

The coefficient of z^r is the binomial coefficient and is denoted by $\binom{n}{r} = {}_nC_r$:

$$(x+y)^n = \sum_{r=0}^{n} \binom{n}{r} x^{n-r} y^r = \sum_{r=0}^{n} \frac{n!}{(n-r)!r!} x^{n-r} y^r.$$

The summation terminates with the term for $r = n$. If n is other than a positive integer the series is infinite, and it converges only if $z = |y/x|$ is less than 1; for $n > 0$ the series is also convergent for $z = 1$.

$$(1+x)^{\frac{1}{2}} = 1 + \frac{1}{2}x - \frac{1 \cdot 1}{2 \cdot 4}x^2 + \frac{1 \cdot 1 \cdot 3}{2 \cdot 4 \cdot 6}x^3 - \frac{1 \cdot 1 \cdot 3 \cdot 5}{2 \cdot 4 \cdot 6 \cdot 8}x^5 + \cdots,$$

$$(1+x)^{-\frac{1}{2}} = 1 - \frac{1}{2}x + \frac{1 \cdot 3}{2 \cdot 4}x^2 - \frac{1 \cdot 3 \cdot 5}{2 \cdot 4 \cdot 6}x^3 + \frac{1 \cdot 3 \cdot 5 \cdot 7}{2 \cdot 4 \cdot 6 \cdot 8}x^4 - \cdots,$$

$$(1+x)^{\frac{1}{3}} = 1 + \frac{x}{3} - \frac{x^2}{9} + \frac{5x^3}{81} - \frac{10x^4}{243} + \frac{22x^5}{729} - \cdots,$$

$$(1+x)^{-\frac{1}{3}} = 1 - \frac{x}{3} + \frac{2x^2}{9} - \frac{14x^3}{81} + \frac{35x^4}{243} - \frac{91x^5}{729} - \cdots.$$

8.4   Progression and series

Arithmetic Progression: a sequence of terms, each of which is the sum of the preceding term and a fixed constant; a, $a + d$, $a + 2d$, $a + 3d$, ... The sums of n successive terms of an arithmetic progression form an arithmetic series:

$$S_n = a + (a + d) + (a + 2d) + (a + 3d) + \cdots + [a + (n - 1)d]$$
$$= na + \frac{1}{2}n(n - 1)d = \frac{n}{2}[\text{first term} + \text{last term}].$$

Geometric Progression: a sequence of terms, each of which is the product of the preceding term and a fixed constant; a, ar, ar^2, ar^3, ... The sums of n successive terms of a geometric progression form a geometric series:

$$G_n = a + ar + ar^2 + ar^3 + \cdots + ar^{n-1} = a\frac{1 - r^n}{1 - r}.$$

If the absolute value of r is greater than 1, the series is divergent for increasing n; if $|r| < 1$, the series converge to the finite limit $a/(1 - r)$ for $n \to \infty$.

Harmonic Progression: a sequence of terms, each of which is the reciprocal of the terms of an arithmetic progression. Thus

$$\frac{1}{a}, \frac{1}{a + d}, \frac{1}{a + 2d}, \cdots, \frac{1}{a + (n - 1)d}.$$

are in harmonic progression.

8.4.1   Means

Arithmetic Mean of n Quantities:

$$m_{\mathrm{a}} = \frac{1}{n}(a_1 + a_2 + a_3 + \cdots + a_n).$$

Geometric Mean of n Quantities:

$$m_{\mathrm{g}} = (a_1 a_2 a_3 \cdots a_n)^{1/n}.$$

Harmonic Mean of n Quantities:

$$m_{\mathrm{h}} = \frac{n}{\dfrac{1}{a_1} + \dfrac{1}{a_2} + \dfrac{1}{a_3} + \cdots + \dfrac{1}{a_n}}.$$

If $a_j > 0$ for all j, $m_{\mathrm{a}} \geq m_{\mathrm{g}} \geq m_{\mathrm{h}}$. Equality is achieved, $m_{\mathrm{a}} = m_{\mathrm{g}} = m_{\mathrm{h}}$, if and only if $a_1 = a_2 = \cdots = a_n$.

8.4.2   Summation formulas

$$1 + 2 + 3 + \cdots + n = \frac{1}{2}n(n+1),$$

$$1^2 + 2^2 + 3^2 + \cdots + n^2 = \frac{1}{6}n(n+1)(2n+1),$$

$$1^3 + 2^3 + 3^3 + \cdots + n^3 = \frac{1}{4}n^2(n+1)^2,$$

$$1^4 + 2^4 + 3^4 + \cdots + n^4 = \frac{1}{30}n(n+1)(2n+1)(3n^2+3n+1).$$

8.5   Taylor series

$$f(x+h) =$$
$$f(x) + hf'(x) + \frac{h^2}{2!}f''(x) + \frac{h^3}{3!}f'''(x) + \cdots + \frac{h^{n-1}}{(n-1)!}f^{(n-1)}(x) + R_n,$$

where, for a suitable choice of θ, $0 < \theta < 1$,

$$R_n = \frac{h^n}{n!}f^{(n)}(x+\theta h) \qquad \text{(Lagrange)},$$

$$= \frac{h^n(1-\theta)^{n-1}}{(n-1)!}f^{(n)}(x+\theta h) \qquad \text{(Cauchy)},$$

$$= \frac{h^n(1-\theta)^{n-p}}{p(n-1)!}f^{(n)}(x+\theta h) \qquad \text{(Schloemilch)} .$$

The Schloemilch expression for R_n reduces to the Lagrange expression for $p = n$, and to the Cauchy expression for $p = 1$.

$$f(x+h, y+k) = f(x,y) + \left(h\frac{\partial f(x,y)}{\partial x} + k\frac{\partial f(x,y)}{\partial y} \right)$$
$$+ \frac{1}{2!}\left(h^2\frac{\partial^2 f(x,y)}{\partial x^2} + 2hk\frac{\partial^2 f(x,y)}{\partial x \partial y} + k^2\frac{\partial^2 f(x,y)}{\partial y^2} \right)$$
$$+ \frac{1}{3!}\left(h^3\frac{\partial^3 f(x,y)}{\partial x^3} + 3h^2k\frac{\partial^3 f(x,y)}{\partial x^2 \partial y} \right.$$
$$\left. + 3hk^2\frac{\partial^3 f(x,y)}{\partial x \partial y^2} + k^3\frac{\partial^3 f(x,y)}{\partial y^3} \right) + \cdots.$$

8.6   Differentiation

a, c, n are constants; u, v, w are variables

$$\frac{d}{dx}(au) = a\frac{du}{dx},$$

$$\frac{d}{dx}(u+v) = \frac{du}{dx} + \frac{dv}{dx},$$

$$\frac{d(uv)}{dx} = \frac{du}{dx}v + u\frac{dv}{dx},$$

$$\frac{d^2(uv)}{dx^2} = \frac{d^2u}{dx^2}v + 2\frac{du}{dx}\frac{dv}{dx} + u\frac{d^2v}{dx^2},$$

$$\frac{du^n}{dx} = nu^{n-1}\frac{du}{dx},$$

$$\frac{d(u/v)}{dx} = \frac{1}{v^2}\left[\frac{du}{dx}v - u\frac{dv}{dx}\right],$$

$$\frac{du(w)}{dx} = \frac{du}{dw}\frac{dw}{dx},$$

$$\frac{du^v}{dx} = u^{v-1}v\frac{du}{dx} + u^v \ln u\frac{dv}{dx},$$

$$\frac{d^n(uv)}{dx^n} = \frac{d^nu}{dx^n}v + n\frac{d^{n-1}u}{dx^{n-1}}\frac{dv}{dx} + \cdots + \binom{n}{k}\frac{d^{n-k}u}{dx^{n-k}}\frac{d^kv}{dx^k} + \cdots,$$

$$\frac{dx^n}{dx} = nx^{n-1} \quad \frac{dx^{\frac{1}{2}}}{dx} = \frac{1}{2}x^{-\frac{1}{2}} \quad \frac{d}{dx}\frac{1}{x} = -\frac{1}{x^2}.$$

8.7   Integration

$$\int f(u)\,dx = \int \frac{f(u)}{\dfrac{du}{dx}}\,du = \int f(u)\frac{dx}{du}\,du,$$

$$\int f(a+bx)\,dx = \frac{1}{b}\int f(y)\,dy,$$

$$\int f(\sqrt{a+bx})\,dx = \frac{2}{b}\int f(y)y\,dy,$$

Integration by Parts:

$$\int u\,dv = uv - \int v\,du, \qquad \int u\frac{dv}{dx}\,dx = uv - \int \frac{du}{dx}v\,dx,$$

$$\int uv\,dx = \left(\int u\,dx\right)v - \int\left(\int u, dx\right)\frac{dv}{dx}\,dx,$$

$$\int x^a\,dx = \begin{cases} \dfrac{x^{a+1}}{a+1} + c & (a \neq -1) \\[2ex] \ln cx & (a = -1) \end{cases}$$

$$\int dx = x + c, \qquad \int x\,dx = \frac{1}{2}x^2 + c,$$

$$\int \frac{dx}{1+x^2} = \arctan x + c,$$

$$\int \frac{dx}{a^2 + b^2 x^2} = \frac{1}{ab}\arctan \frac{bx}{a} + c,$$

$$\int \frac{dx}{a^2 - b^2 x^2} = \frac{1}{2ab}\ln \frac{a+bx}{a-bx} + c,$$

$$\int \frac{dx}{\sqrt{a^2 + b^2 x^2}} = \frac{1}{b}\operatorname{arsinh} \frac{bx}{a} + c,$$

$$= \frac{1}{b}\ln\left[bx + \sqrt{a^2 + b^2 x^2}\right] + c,$$

$$\int \frac{dx}{\sqrt{a^2 - b^2 x^2}} = \frac{1}{b}\arcsin \frac{bx}{a} + c,$$

$$\int \frac{dx}{\sqrt{b^2 x^2 - a^2}} = \frac{1}{b}\ln\left[bx + \sqrt{b^2 x^2 - a^2}\right] + c,$$

$$\int \sqrt{a^2 + b^2 x^2}\,dx = \frac{1}{2}\left[x\sqrt{a^2 + b^2 x^2} + a^2 \int \frac{dx}{\sqrt{a^2 + b^2 x^2}}\right].$$

8.8 Logarithmic functions

$$\log_a b \log_b c = \log_a c, \qquad \log_b a = 1/\log_a b,$$

$$\log_a x = \frac{\log_b x}{\log_b a} = \frac{\ln x}{\ln a},$$

$$\ln x \equiv \log_e x = (\log_e 10)\log_{10} x, \qquad \ln 10 = \log_e 10 = 2.302\,585\,092\ldots$$

Series:

$$\ln(1+x) = x - \frac{x^2}{2} + \frac{x^3}{3} - \frac{x^4}{4} + \cdots \qquad (-1 < x \leq 1)$$

$$\ln x = \ln a + \frac{x-a}{a} - \frac{(x-a)^2}{2a^2} + \frac{(x-a)^3}{3a^3} - \cdots \qquad (0 < x \leq 2a)$$

$$\ln \frac{1+x}{1-x} = 2x \left[1 + \frac{x^2}{3} + \frac{x^4}{5} + \frac{x^6}{7} + \cdots \right] \qquad (x^2 < 1),$$

$$\ln x = 2 \frac{x-1}{x+1} \left[1 + \frac{(x-1)^2}{3(x+1)^2} + \frac{(x-1)^4}{5(x+1)^4} + \cdots \right] \qquad (x > 0)$$

$$= \frac{x^2-1}{x^2+1} \left[1 + \frac{(x^2-1)^2}{3(x^2+1)^2} + \frac{(x^2-1)^4}{5(x^2+1)^4} + \cdots \right] \qquad (x > 0).$$

Euler's Constant:

$$\gamma = \lim_{m \to \infty} \left[1 + \frac{1}{2} + \frac{1}{3} + \cdots + \frac{1}{m} - \ln m \right] = 0.577\,215\,6649 \ldots.$$

Derivatives:

$$\frac{d \ln x}{dx} = \frac{1}{x},$$

$$\frac{d}{dx} \ln g(x) = \frac{1}{g(x)} \frac{dg(x)}{dx}.$$

Integrals:

$$\int \ln x \, dx = x \ln x - x + c,$$

$$\int x^a \ln x \, dx = \frac{x^{a+1}}{a+1} \left[\ln x - \frac{1}{a-1} \right] + c.$$

8.9   Exponential functions

$$e^x = \exp x = 1 + \frac{x}{1!} + \frac{x^2}{2!} + \frac{x^3}{3!} + \cdots + \frac{x^n}{n!} + \cdots$$

$$= \lim_{n \to \infty} \left[1 + \frac{x}{n} \right]^{n+1/2x}.$$

Derivatives:

$$\frac{de^x}{dx} = e^x, \qquad\qquad \frac{d}{dx} e^{u(x)} = e^{u(x)} \frac{du}{dx},$$

$$\frac{da^x}{dx} = a^x \ln a, \qquad\qquad \frac{dx^x}{dx} = (1 + \ln x)x^x,$$

$$\frac{du^v}{dx} = \left[\frac{dv}{dx} \ln u + \frac{v}{u} \frac{du}{dx} \right].$$

Integrals:

$$\int e^{ax}\,dx = \frac{e^{ax}}{a} + c,$$

$$\int xe^{ax}\,dx = \frac{e^{ax}}{a^2}(ax - 1) + c,$$

$$\int x^2 e^{ax}\,dx = 2!\frac{e^{ax}}{a^3}\left(\frac{a^2x^2}{2} - ax + 1\right) + c,$$

$$\int x^3 e^{ax}\,dx = 3!\frac{e^{ax}}{a^4}\left(\frac{a^3x^3}{3!} - \frac{a^2x^2}{2!} + ax - 1\right) + c,$$

$$\int \frac{e^{ax}}{x^n}\,dx = -\frac{e^{ax}}{(n-1)x^{n-1}} + \frac{a}{n-1}\int \frac{e^{ax}}{x^{n-1}}dx + c.$$

8.10   Trigonometric functions

$$\sin^2 A + \cos^2 A = 1,$$
$$\sin(A + B) = \sin A \cos B + \cos A \sin B,$$
$$\cos(A + B) = \cos A \cos B - \sin A \sin B,$$
$$2\sin A \sin B = \cos(A - B) - \cos(A + B),$$
$$2\sin A \cos B = \sin(A + B) + \sin(A - B),$$
$$2\cos A \cos B = \cos(A + B) + \cos(A - B),$$
$$\sin 2A = 2\sin A \cos A,$$
$$\cos 2A = \cos^2 A - \sin^2 A = 1 - 2\sin^2 A = 2\cos^2 A - 1,$$
$$(\cos x + i\sin x)^\nu = \cos\nu x + i\sin\nu x \qquad \text{(DeMoivre's Theorem)},$$
$$e^{i\pi} + 1 = 0 \qquad\qquad\qquad \text{(Euler's Formula)},$$
$$\cos x = \frac{1}{2}(e^{ix} + e^{-ix}),$$
$$\sin x = \frac{1}{2i}(e^{ix} - e^{-ix}).$$

Series Expansions:

$$\sin z = z - \frac{z^3}{3!} + \frac{z^5}{5!} - \frac{z^7}{7!} + \frac{z^9}{9!} - \cdots,$$

$$\cos z = 1 - \frac{z^2}{2!} + \frac{z^4}{4!} - \frac{z^6}{6!} + \frac{z^8}{8!} - \cdots,$$

$$\tan z = z + \frac{z^3}{3} + \frac{2z^5}{15} + \frac{17z^7}{315} + \frac{62z^9}{2835} + \cdots, \qquad |z| < \frac{\pi}{2},$$

$$\cot z = \frac{1}{z} - \frac{z}{3} - \frac{z^3}{45} - \frac{2z^5}{945} - \frac{z^7}{4725} + \cdots, \qquad |z| < \pi,$$

$$\sec z = 1 + \frac{z^2}{2} + \frac{5z^4}{24} + \frac{61z^6}{720} + \frac{277z^8}{8064} + \cdots, \qquad |z| < \frac{\pi}{2},$$

$$\csc z = \frac{1}{z} + \frac{z}{6} + \frac{7z^3}{360} + \frac{31z^5}{15120} + \frac{127z^7}{604800} + \cdots, \qquad |z| < \pi,$$

$$e^{\sin z} = 1 + z + \frac{z^2}{2} - \frac{z^4}{8} - \frac{z^5}{15} - \frac{z^6}{240} + \frac{z^7}{90} + \frac{31z^8}{5760} + \cdots.$$

Derivatives:

$$\frac{d}{dx}\sin x = \cos x, \qquad \frac{d}{dx}\cot x = -\csc^2 x,$$

$$\frac{d}{dx}\cos x = -\sin x, \qquad \frac{d}{dx}\sec x = \sec x \tan x,$$

$$\frac{d}{dx}\tan x = \sec^2 x, \qquad \frac{d}{dx}\csc x = -\csc x \cot x.$$

Integrals:

$$\int \sin x \, dx = -\cos x + c,$$

$$\int x \sin x \, dx = \sin x - x \cos x + c,$$

$$\int \sin^2 x \, dx = \frac{x}{2} - \frac{\sin 2x}{4} + c = \frac{1}{2}x - \sin x \cos x + c,$$

$$\int \cos x \, dx = \sin x + c,$$

$$\int x \cos x \, dx = \cos x + x \sin x + c,$$

$$\int \cos x \sin x \, dx = \sin^2 x + c,$$

$$\int \cos^2 x \, dx = x - \int \sin^2 x \, dx,$$

$$\int \tan x \, dx = \ln \sec x + c = -\ln \cos x + c,$$

$$\int \tan^2 x \, dx = \tan x - x + c,$$

$$\int \cot x \, dx = \ln[c \sin x],$$

$$\int \sec x \, dx = \ln[c(\sec x + \tan x)] = \ln\left[c\tan\left(\frac{\pi}{4} + \frac{x}{2}\right)\right],$$

$$\int \csc x \, dx = \ln\left[c\tan\left(\frac{1}{2}x\right)\right] = \ln[c(\csc x - \cot x)].$$

Plane Triangles:

Let a, b, and c be the sides opposite the angles A, B, and C, respectively,

$$A + B + C = 180° = \pi \text{ rad},$$
$$\frac{\sin A}{a} = \frac{\sin B}{b} = \frac{\sin C}{c},$$
$$a = b\cos C + c\cos B,$$
$$b = c\cos A + a\cos C,$$
$$c = a\cos B + b\cos A,$$
$$c^2 = a^2 - 2ab\cos C + b^2, \qquad \cos C = \frac{a^2 + b^2 - c^2}{2ab},$$
$$\frac{a+b}{a-b} = \frac{\tan\frac{1}{2}(A+B)}{\tan\frac{1}{2}(A-B)}.$$

Area of a Triangle:

$$S = \tfrac{1}{2}ab\sin C = \frac{1}{2}a^2\frac{\sin B \sin C}{\sin A}$$
$$= \tfrac{1}{4}\sqrt{[(a+b)^2 - c^2][c^2 - (a-b)^2]} = \sqrt{s(s-a)(s-b)(s-c)},$$

where $s = (a + b + c)/2$ is the semi-perimeter.

Spherical Triangles:

On the unit sphere, let a, b, and c be the sides opposite the angles A, B, and C, respectively.

$$A + B + C > \pi, \qquad \text{spherical excess, } E = A + B + C - \pi,$$
$$\text{area of a sphere: } S = 4\pi, \qquad \text{area of a spherical triangle: } S = E,$$

$$\frac{\sin A}{\sin a} = \frac{\sin B}{\sin b} = \frac{\sin C}{\sin c},$$
$$\cos c = \cos a \cos b + \sin a \sin b \cos C = \frac{\cos a \cos(b - \theta)}{\cos \theta},$$

where $\tan \theta = \tan a \cos C$,

$$\cos C = -\cos A \cos B + \sin A \sin B \cos c.$$

8.11   Inverse trigonometric functions

Series Expansions:

$$\arcsin x = x + \frac{x^3}{2\cdot 3} + \frac{1\cdot 3\,x^5}{2\cdot 4\cdot 5} + \cdots + \frac{(2k-1)!!\,x^{2k+1}}{(2k)!!(2k+1)} + \cdots,$$

$$\arctan x = \begin{cases} x - \dfrac{x^3}{3} + \dfrac{x^5}{5} - \dfrac{x^7}{7} + \cdots, & x^2 < 1 \\[2mm] \dfrac{\pi}{2} - \dfrac{1}{x} + \dfrac{1}{3x^3} - \dfrac{1}{5x^5} + \cdots, & x^2 > 1 \end{cases}.$$

Derivatives:

$$\frac{d}{dx}\arcsin\frac{x}{a} = \frac{1}{\sqrt{a^2 - x^2}},$$

$$\frac{d}{dx}\arccos\frac{x}{a} = \frac{-1}{\sqrt{a^2 - x^2}},$$

$$\frac{d}{dx}\arctan\frac{x}{a} = \frac{a}{a^2 + x^2},$$

$$\frac{d}{dx}\operatorname{arccot}\frac{x}{a} = \frac{-a}{a^2 + x^2},$$

$$\frac{d}{dx}\operatorname{arcsec}\frac{x}{a} = \frac{a}{x\sqrt{x^2 - a^2}},$$

$$\frac{d}{dx}\operatorname{arccsc}\frac{x}{a} = \frac{-a}{x\sqrt{x^2 - a^2}}.$$

Integrals:

$$\int \arcsin\frac{x}{a}\,dx = x\arcsin\frac{x}{a} + \sqrt{a^2 - x^2} + c,$$

$$\int \arccos\frac{x}{a}\,dx = x\arccos\frac{x}{a} - \sqrt{a^2 - x^2} + c,$$

$$\int \arctan\frac{x}{a}\,dx = x\arctan\frac{x}{a} - \frac{a}{2}\ln a^2 + x^2 + c,$$

$$\int \operatorname{arccot}\frac{x}{a}\,dx = \operatorname{arccot}\frac{x}{a} + \frac{a}{2}\ln a^2 + x^2 + c,$$

$$\int \operatorname{arcsec}\frac{x}{a}\,dx = \operatorname{arcsec}\frac{x}{a} - a\ln\left[x + \sqrt{x^2 - a^2}\right] + c$$
$$(0 < \operatorname{arcsec}(x/a) < \pi/2),$$

$$= \operatorname{arcsec}\frac{x}{a} + a\ln\left[x + \sqrt{x^2 - a^2}\right] + c$$
$$(\pi/2 < \operatorname{arcsec}(x/a) < \pi),$$

$$\int \operatorname{arccsc} \frac{x}{a}\, dx = \operatorname{arccsc} \frac{x}{a} + a \ln\left[x + \sqrt{x^2 - a^2}\,\right] + c$$

$$(0 < \operatorname{arccsc}(x/a) < \pi/2),$$

$$= \operatorname{arccsc} \frac{x}{a} - a \ln\left[x + \sqrt{x^2 - a^2}\,\right] + c$$

$$(-\pi/2 < \operatorname{arccsc}(x/a) < 0),$$

$$\int x \arcsin \frac{x}{a}\, dx = \tfrac{1}{4}\left[(2x^2 - a^2)\arcsin \frac{x}{a} + x\sqrt{a^2 - x^2}\,\right] + c,$$

$$\int x \arctan \frac{x}{a}\, dx = \tfrac{1}{2}\left[(a^2 + x^2)\arctan \frac{x}{a} - ax\right] + c.$$

8.12 Hyperbolic functions

Each trigonometric relation will yield a corresponding hyperbolic relation by the replacements

$$x \to ix, \quad \cos x \to \cosh x, \quad \text{and} \quad \sin x \to i \sinh x.$$

$$\sin ix = i \sinh x, \qquad \cos ix = \cosh x, \qquad \tan ix = i \tanh x,$$
$$\csc ix = -i \operatorname{csch} x, \qquad \sec ix = \operatorname{sech} x, \qquad \cot ix = -i \coth x,$$
$$\sin(x + iy) = \sin x \cosh y + i \cos x \sinh y,$$
$$\cos(x + iy) = \cos x \cosh y - i \sin x \sinh y,$$
$$\cosh^2 x + \sinh^2 x = 1,$$
$$\cosh x + \sinh x = e^x, \qquad\qquad \cosh x = \tfrac{1}{2}(e^x + e^{-x}),$$
$$\cosh x - \sinh x = e^{-x}, \qquad\qquad \sinh x = \tfrac{1}{2}(e^x - e^{-x}),$$
$$\sinh(A + B) = \sinh A \cosh B + \cosh A \sinh B,$$
$$\cosh(A + B) = \cosh A \cosh B + \sinh A \sinh B,$$
$$2 \sinh A \sinh B = \cosh(A + B) - \cosh(A - B),$$
$$2 \sinh A \cosh B = \sinh(A + B) + \sinh(A_B),$$
$$2 \cosh A \cosh B = \cosh(A + B) + \cosh(A + B).$$

Series Expansions:

$$\cosh x = 1 + \frac{x^2}{2!} + \frac{x^4}{4!} + \frac{x^6}{6!} + \frac{x^8}{8!} + \cdots,$$

$$\sinh x = x + \frac{x^3}{3!} + \frac{x^5}{5!} + \frac{x^7}{7!} + \frac{x^9}{9!} + \cdots,$$

$$\tanh x = x - \frac{x^3}{3} + \frac{2x^5}{15} - \frac{17x^7}{315} + \cdots,$$

$$\coth x = \frac{1}{x} + \frac{x}{3} - \frac{x^3}{45} + \frac{2x^5}{945} + \cdots,$$

$$\operatorname{sech} x = 1 - \frac{x^2}{2} + \frac{5x^4}{24} - \frac{61x^6}{720} + \cdots,$$

$$\operatorname{csch} x = \frac{1}{x} - \frac{x}{6} + \frac{7x^3}{360} - \frac{31x^5}{1520} + \frac{127x^7}{423360} + \cdots.$$

Derivatives:

$$\frac{d}{dx}\sinh x = \cosh x, \qquad \frac{d}{dx}\coth x = -\operatorname{csch}^2 x,$$

$$\frac{d}{dx}\cosh x = \sinh x, \qquad \frac{d}{dx}\operatorname{sech} x = -\operatorname{sech} x \tanh x,$$

$$\frac{d}{dx}\tanh x = \operatorname{sech}^2 x, \qquad \frac{d}{dx}\operatorname{csch} x = -\operatorname{csch} x \coth x.$$

Integrals:

$$\int \sinh x \, dx = \cosh x + c,$$

$$\int x \sinh x \, dx = x \cosh x - \sinh x + c,$$

$$\int \cosh x \, dx = \sinh x + c,$$

$$\int x \cosh x \, dx = x \sinh x - \cosh x + c,$$

$$\int \cosh x \sinh x \, dx = \frac{1}{2}\sinh^2 x + c,$$

$$\int \tanh x \, dx = \ln \cosh x + c,$$

$$\int \tanh^2 x \, dx = x - \tanh x + c,$$

$$\int \cot x \, dx = \ln \sinh x + c,$$

$$\int \operatorname{sech} x \, dx = \arctan \sinh x + c,$$

$$\int \operatorname{csch} x \, dx = \ln\left[c\tanh\frac{x}{2}\right] = \ln\left[c\tan\left(\frac{\pi}{4} + \frac{x}{2}\right)\right].$$

8.13   Inverse hyperbolic functions

$$\operatorname{arsinh} x = \int_0^x \frac{dt}{(1+t^2)^{1/2}} = \ln\left[x + (x^2+1)^{1/2}\right],$$

$$\operatorname{arcosh} x = \int_0^x \frac{dt}{(t^2-1)^{1/2}} = \ln\left[x + (x^2-1)^{1/2}\right],$$

$$\operatorname{artanh} x = \int_0^x \frac{dt}{1-t^2} = \frac{1}{2}\ln\frac{1+x}{1-x} = \ln\left[x + (x^2+1)^{1/2}\right],$$

Series Expansions:

$$\operatorname{arsinh} x = \begin{cases} x - \dfrac{x^3}{2\cdot3} + \dfrac{1\cdot3x^5}{2\cdot4\cdot5} + \dfrac{1\cdot3\cdot5x^7}{2\cdot4\cdot6\cdot7} + \cdots, & |x| < 1, \\[2ex] \ln 2x + \dfrac{1}{2\cdot2x^2} - \dfrac{1\cdot3}{2\cdot4\cdot4x^4} + \dfrac{1\cdot3\cdot5}{2\cdot4\cdot6\cdot6x^6} + \cdots, & |x| > 1, \end{cases}$$

$$\operatorname{arcosh} x = \ln 2x - \frac{1}{2\cdot2x^2} - \frac{1\cdot3}{2\cdot4\cdot4x^4} - \frac{1\cdot3\cdot5}{2\cdot4\cdot6\cdot6x^6} - \cdots, \quad (|x| > 1),$$

$$\operatorname{artanh} x = \operatorname{arcoth}\frac{1}{x} = x + \frac{x^3}{3} + \frac{x^5}{5} + \frac{x^7}{7} + \cdots, \quad (|x| < 1),$$

Derivatives:

$$\frac{d}{dx}\operatorname{arsinh}\frac{x}{a} = (a^2 + x^2)^{-1/2},$$

$$\frac{d}{dx}\operatorname{arcosh}\frac{x}{a} = (x^2 - a^2)^{-1/2},$$

$$\frac{d}{dx}\operatorname{arctan}\frac{x}{a} = a/(a^2 - x^2),$$

$$\frac{d}{dx}\operatorname{arcoth}\frac{x}{a} = -a/(x^2 - a^2),$$

$$\frac{d}{dx}\operatorname{arsech}\frac{x}{a} = -\frac{a}{x(a^2 - x^2)^{1/2}}, \quad x \neq 0,$$

$$\frac{d}{dx}\operatorname{arcsch}\frac{x}{a} = -\frac{a}{|x|(a^2 + x^2)^{1/2}}, \quad x \neq 0,$$

Integrals:

$$\int \operatorname{arsinh}\frac{x}{a}\,dx = x\operatorname{arsinh}\frac{x}{a} - \sqrt{a^2 + x^2} + c,$$

$$\int \operatorname{arcosh}\frac{x}{a}\,dx = x\operatorname{arcosh}\frac{x}{a} - \sqrt{x^2 - a^2} + c,$$

$$\int \operatorname{artanh}\frac{x}{a}\,dx = x\operatorname{artanh}\frac{x}{a} + \frac{a}{2}\ln(a^2 - x^2) + c,$$

$$\int \operatorname{arcoth} \frac{x}{a}\, dx = x \operatorname{arcoth} \frac{x}{a} + \frac{a}{2} \ln(x^2 - a^2) + c,$$

$$\int \operatorname{arsech} \frac{x}{a}\, dx = x \operatorname{arsech} \frac{x}{a} + a \operatorname{arsinh} \frac{x}{a} + c,$$

$$\int \operatorname{arcsch} \frac{x}{a}\, dx = x \operatorname{arcsch} \frac{x}{a} + a \operatorname{arcsch} \frac{x}{a} + c,$$

$$\int x \operatorname{arsinh} \frac{x}{a}\, dx = \tfrac{1}{4}\left[(2x^2 + a^2) \operatorname{arsinh} \frac{x}{a} - x\sqrt{x^2 + a^2}\right] + c,$$

$$\int x \operatorname{artanh} \frac{x}{a}\, dx = \tfrac{1}{2}\left[(x^2 - a^2) \arctan \frac{x}{a} + ax\right] + c.$$

8.14 Gamma function

The gamma function is the analytic continuation into the complex plane of the factorial $n!$

$$\Gamma(z) = \int_0^{\infty} t^{z-1} e^{-t} dt, \quad \Re(z) > 0.$$

$$\Gamma(z+1) = z\Gamma(z), \qquad \Gamma(1) = 1, \qquad \Gamma\left(\frac{1}{2}\right) = \sqrt{\pi}.$$

For n a positive integer, $\Gamma(n) = (n-1)!$

$$\Gamma\left(n + \frac{1}{2}\right) \equiv \left(n - \frac{1}{2}\right)! = \frac{(2n-1)!!}{2^n}\sqrt{\pi} = \frac{(2n)!}{2^{2n}n!}\sqrt{\pi},$$

$$\Gamma(z)\Gamma(1-z) = \frac{\pi}{\sin \pi z}, \qquad \Gamma(z)\Gamma(-z) = -\frac{\pi}{z \sin \pi z},$$

$$\Gamma\left(1 + \frac{z}{\pi}\right)\Gamma\left(1 - \frac{z}{\pi}\right) = \frac{z}{\sin z},$$

$$\Gamma(2z) = \frac{2^{2z-1}\Gamma(z)\Gamma\left(z + \frac{1}{2}\right)}{\Gamma\left(\frac{1}{2}\right)},$$

$$\Gamma(z) \sim \sqrt{2\pi}\, z^{z-1/2} \exp\left[-z + \frac{1}{12z} - \frac{1}{360z^3} + \frac{1}{1260z^5} - \frac{1}{1680z^7} + \cdots\right],$$

$$\sim \sqrt{2\pi}\, z^{z-1/2} e^{-z}\left[1 + \frac{1}{12z} + \frac{1}{288z^2} - \frac{139}{51840z^3} - \frac{571}{2488320z^4} + \cdots\right],$$

$$|z| \to \infty, \quad |\arg z| < \pi,$$

$$n! \sim \sqrt{2\pi n}\, n^n \exp\left[-n + \frac{1}{12n} - \frac{1}{360n^3} + \frac{1}{1260n^5} - \frac{1}{1680n^7} + \cdots\right],$$

$$\sim \sqrt{2\pi n}\, n^n e^{-n} \left[1 + \frac{1}{12n} + \frac{1}{288n^2} - \frac{139}{51840n^3} - \frac{571}{2488320n^4} + \cdots \right],$$

$$\Gamma(x+a) \approx \left(x + \frac{a-1}{2} \right)^a \Gamma(x).$$

8.15 Definite integrals

$$\int_1^\infty \frac{dx}{x^m} = \frac{1}{m-1},$$

$$\int_0^a \frac{dx}{\sqrt{a^2 - x^2}} = \frac{\pi}{2},$$

$$\int_0^\infty \frac{y\, dx}{y^2 + x^2} = \begin{cases} -\frac{\pi}{2}, & y < 0 \\ 0, & y = 0 \\ \frac{\pi}{2}, & y > 0 \end{cases},$$

$$\frac{2}{\pi} \int_0^{\pi/2} \sin^{2n} x\, dx = \frac{2}{\pi} \int_0^{\pi/2} \cos^{2n} x\, dx = \frac{(2n-1)!!}{(2n)!!}$$

$$= \frac{(2n)!}{(2^n n!)^2} = \frac{1}{2^{2n}} \binom{2n}{n} = \frac{\Gamma(n + \frac{1}{2})}{\Gamma(\frac{1}{2})\Gamma(n+1)},$$

$$\frac{2}{\pi} \int_0^\pi \sin^{2n} x \cos^{2m} x\, dx = \frac{(2n)!(2m)!}{2^{2n+2m} n! m! (n+m)!},$$

$$2 \int_0^{\pi/2} \sin^{2z-1} x \cos^{2w-1} x\, dx = \int_0^1 t^{z-1}(1-t)^{w-1}\, dt = \frac{\Gamma(z)\Gamma(w)}{\Gamma(z+w)},$$

$$\int_0^x \frac{\sin t}{t}\, dt = \operatorname{Si}(x) = x - \frac{x^3}{3 \cdot 3!} + \frac{x^5}{5 \cdot 5!} - \frac{x^7}{7 \cdot 7!} + \cdots,$$

$$\int_0^x \frac{1 - \cos t}{t}\, dt = \operatorname{Cin}(x) = \frac{x^2}{2 \cdot 2!} - \frac{x^4}{4 \cdot 4!} + \frac{x^6}{6 \cdot 6!} - \cdots$$

$$= \ln x + \gamma - \operatorname{Ci}(x),$$

$$\gamma = 0.577\,215\,6649\ldots \quad \text{(Euler's constant)},$$

$$\int_1^\infty \frac{e^{-xt}}{t^n}\, dt = E_n(x),$$

$$E_{n+1}(x) = \frac{1}{n}[e^{-x} - xE_n(x)], \quad E_0(x) = \frac{e^{-x}}{x},$$

$$E_1(x) = -\ln x - \gamma + x - \frac{x^2}{2 \cdot 2!} + \frac{x^3}{3 \cdot 3!} - \frac{x^4}{4 \cdot 4!} + \cdots,$$

$$\int_0^\infty e^{-a^2 x^2}\, dx = \frac{\sqrt{\pi}}{2a},$$

$$\int_0^\infty x e^{-a^2 x^2}\, dx = \frac{1}{2a^2},$$

$$\int_0^\infty x^2 e^{-a^2 x^2}\, dx = \frac{\sqrt{\pi}}{4a^3},$$

$$\int_0^\infty x^{2n} e^{-px^2}\, dx = \frac{(2n-1)!!}{2^{n+1}p^n}\sqrt{\frac{\pi}{p}}, \quad n \text{ integer},$$

$$\int_0^\infty x^{2n+1} e^{-px^2}\, dx = \frac{n!}{2p^{n+1}}, \quad n \text{ integer}.$$

8.16   Delta function

The discrete delta function, or Kronecker symbol, is the quantity

$$\delta_{ij} = \begin{cases} 1 & \text{if } i = j \\ 0 & \text{if } i \neq j \end{cases}.$$

The Dirac delta function $\delta(x)$ is an improper function (distribution) defined such that

$$x\delta(x) = 0 \quad \text{and} \quad \int_a^b \delta(x)\, dx = \begin{cases} 1 & \text{if } a < 0 < b \\ 0 & \text{otherwise} \end{cases}.$$

For an arbitrary function $f(x)$, and for $a < y < b$,

$$\int_a^b f(x)\delta(x - y)dx = f(y),$$

$$\int_a^b f(x)\delta'(x - y)dx = -f'(y),$$

where the prime denotes differentiation with respect to the argument.

If the δ function has as an argument a function $g(x)$ of the independent variable x with simple zeros at x_i [$g(x_i) = 0$, $g'(x_i) \neq 0$], it can be transformed according to the rule

$$\delta(g(x)) = \sum_i \frac{1}{|g'(x_i)|} \delta(x - x_i).$$

8.17   Vector algebra

A, B, y, z, etc., are scalars; $\mathbf{A}$, $\mathbf{B}$, $\mathbf{y}$, $\mathbf{z}$, etc., are vectors. In particular, a vector and its scalar magnitude are denoted by the same symbol, with the vector in bold type: $X = |\mathbf{X}| = \sqrt{\mathbf{X} \cdot \mathbf{X}}$.

[Mathematical vectors are usually denoted by upright bold-face type but *physical quantities* expressed as three-dimensional vectors (in accord with International Organization for Standardization recommendations[4]) preferably should be set in *slanted* boldface type.]

8.17.1   Dot product (scalar product or inner product)

$$\mathbf{A} \cdot \mathbf{B} = (\mathbf{A}, \mathbf{B}) = AB \cos\phi,$$

ϕ being the angle between $\mathbf{A}$ and $\mathbf{B}$.

In an orthogonal coordinate system defined by unit vectors $\mathbf{e}_i$, ($i = 1, 2, 3$), $\mathbf{e}_i \cdot \mathbf{e}_j = \delta_{ij}$,

$$\mathbf{A} = A_1 \mathbf{e}_1 + A_2 \mathbf{e}_2 + A_3 \mathbf{e}_3,$$
$$\mathbf{B} = B_1 \mathbf{e}_1 + B_2 \mathbf{e}_2 + B_3 \mathbf{e}_3,$$

the scalar product is

$$\mathbf{A} \cdot \mathbf{B} = A_1 B_1 + A_2 B_2 + A_3 B_3.$$

8.17.2   Cross product (vector product or outer product)

In a three-dimensional space $\mathbf{C} = \mathbf{A} \times \mathbf{B}$ is a (pseudo-)vector perpendicular to the plane determined by $\mathbf{A}$ and $\mathbf{B}$ whose absolute value is $C = AB \sin\phi$ and whose direction is such that $\mathbf{A}$, $\mathbf{B}$, and $\mathbf{C}$ form a right-handed screw. In an orthogonal coordinate system the cross product is expressed by the determinant

$$\mathbf{C} = \begin{vmatrix} \mathbf{e}_1 & \mathbf{e}_2 & \mathbf{e}_3 \\ A_1 & A_2 & A_3 \\ B_1 & B_2 & B_3 \end{vmatrix}.$$

This can be more succinctly expressed using the totally antisymmetric symbol ϵ_{ijk}: $\epsilon_{123} = 1$, and ϵ_{ijk} changes sign if any two indices are exchanged:

$$\epsilon_{123} = 1, \qquad \epsilon_{ijk} = -\epsilon_{ikj} = -\epsilon_{kji} = -\epsilon_{jik}$$

so that ϵ_{ijk} is zero if any two indices are the same, $+1$ if the indices are an even permutation of 123 and -1 if they are an odd permutation. Then

$$C_i = \sum_{jk} \epsilon_{ijk} A_j B_k.$$

In a two-dimensional space, the cross product is a (pseudo-)scalar,

$$C = A_1 B_2 - A_2 B_1.$$

8.17.3 Vector identities

$$\mathbf{A} \cdot \mathbf{B} = \mathbf{B} \cdot \mathbf{A}, \qquad \mathbf{A} \times \mathbf{B} = -\mathbf{B} \times \mathbf{A},$$
$$[\mathbf{ABC}] \equiv \mathbf{A} \cdot (\mathbf{B} \times \mathbf{C}) = (\mathbf{A} \times \mathbf{B}) \cdot \mathbf{C},$$
$$[\mathbf{ABC}] = [\mathbf{BCA}] = [\mathbf{CAB}] = -[\mathbf{CBA}] = -[\mathbf{BAC}] = -[\mathbf{ACB}],$$

$[\mathbf{ABC}]$ is the volume of the parallelepiped defined by the vectors $\mathbf{A}$, $\mathbf{B}$, $\mathbf{C}$.

$$\mathbf{A} \times (\mathbf{B} \times \mathbf{C}) = (\mathbf{C} \times \mathbf{B}) \times \mathbf{A} = \mathbf{B}(\mathbf{A} \cdot \mathbf{C}) - \mathbf{C}(\mathbf{A} \cdot \mathbf{B}),$$
$$(\mathbf{A} \times \mathbf{B}) \cdot (\mathbf{C} \times \mathbf{D}) = \mathbf{A} \cdot [\mathbf{B} \times (\mathbf{C} \times \mathbf{D})] = (\mathbf{A} \cdot \mathbf{C})(\mathbf{B} \cdot \mathbf{D}) - (\mathbf{A} \cdot \mathbf{D})(\mathbf{B} \cdot \mathbf{C}),$$
$$(\mathbf{A} \times \mathbf{B}) \times (\mathbf{C} \times \mathbf{D}) = \mathbf{C}[\mathbf{ABD}] + \mathbf{D}[\mathbf{CBA}] = \mathbf{A}[\mathbf{CBD}] + \mathbf{B}[\mathbf{ACD}].$$

8.17.4 Vector linear equations

Given $\mathbf{A} \cdot \mathbf{X} = p$, $\mathbf{B} \times \mathbf{X} = \mathbf{C}$, with $\mathbf{B} \cdot \mathbf{C} = 0$:

$$\mathbf{X} = \frac{p\mathbf{B} + \mathbf{C} \times \mathbf{A}}{\mathbf{A} \cdot \mathbf{B}}.$$

Given $\mathbf{A} \cdot \mathbf{X} = p$, $\mathbf{B} \cdot \mathbf{X} = q$, $\mathbf{C} \cdot \mathbf{X} = r$ with $\mathbf{A} \cdot \mathbf{B} \times \mathbf{C} \neq 0$:

$$\mathbf{X} = \frac{p\mathbf{B} \times \mathbf{C} + q\mathbf{C} \times \mathbf{A} + r\mathbf{A} \times \mathbf{B}}{[\mathbf{ABC}]}.$$

8.17.5 Differential operators

The *gradient* of a scalar function $f(\mathbf{r}) = f(x, y, z)$ gives the absolute value and direction of the maximum change in f. The gradient defines a vector field perpendicular to the surfaces $f = \text{const.}$

$$\text{grad}\, f = \nabla f = \lim_{V \to 0} \frac{1}{V} \int_S f \, d\mathbf{S} = \frac{\partial f}{\partial x}\mathbf{i} + \frac{\partial f}{\partial y}\mathbf{j} + \frac{\partial f}{\partial z}\mathbf{k}.$$

The *divergence* of a vector $\mathbf{A}(\mathbf{r}) = \mathbf{A}(x, y, z)$ is the net outflow of the vector flux through the surface of an infinitesimal volume element containing the point $\mathbf{r}$, divided by the volume:

$$\text{div }\mathbf{A} = \nabla\cdot\mathbf{A} = \lim_{V\to 0}\frac{1}{V}\int_S \mathbf{A}\cdot d\mathbf{S} = \frac{\partial A_1}{\partial x_1}\mathbf{e}_1 + \frac{\partial A_2}{\partial x_2}\mathbf{e}_2 + \frac{\partial A_3}{\partial x_3}\mathbf{e}_3.$$

If C is a closed curve in the vector field $\mathbf{A}$, the circulation of $\mathbf{A}$ with respect to C is the contour integral $\oint \mathbf{A}\cdot d\mathbf{s}$. Assigning to the circulation a direction given by the thumb when the fingers of the right hand follow the contour in the positive direction, the *curl* of $\mathbf{A}$ is the vector representing the magnitude and direction of the maximum circulation per unit area, $\mathbf{n}\oint \mathbf{A}\cdot d\mathbf{s}/S$, of a contour enclosing a small area S encircling the field point.

$$\text{curl }\mathbf{A} = \lim_{S\to 0}\frac{\mathbf{n}}{S}\oint_C \mathbf{A}\cdot d\mathbf{s}.$$

For a finite area, one then has

$$\oint_C \mathbf{A}\cdot d\mathbf{s} = \int_S \text{curl }\mathbf{A}\cdot\mathbf{n}\,dS = \int_S \nabla\times\mathbf{A}\cdot\mathbf{n}\,dS = \int_S \nabla\cdot\mathbf{A}\times\mathbf{n}\,dS$$

and

$$\text{curl }\mathbf{A} = \nabla\times\mathbf{A} = \lim_{V\to 0}\frac{1}{V}\int_S \mathbf{n}\times\mathbf{A}\,dS = \begin{vmatrix} \mathbf{e}_1 & \mathbf{e}_2 & \mathbf{e}_3 \\ \dfrac{\partial}{\partial x} & \dfrac{\partial}{\partial y} & \dfrac{\partial}{\partial z} \\ A_1 & A_2 & A_3 \end{vmatrix}$$

The vector differential operator may be represented by

$$\nabla(\cdots) = \lim_{V\to 0}\frac{1}{V}\int_S \mathbf{n}(\cdots)\,dS.$$

in which $(\cdots)$ can be replaced by any valid expression: $f(\mathbf{r})$, $\cdot\mathbf{A}$, $\times\mathbf{B}$, etc.,

$$\nabla(fg) = \nabla(gf) = f\nabla g + g\nabla f,$$

$$\nabla\cdot(f\mathbf{A}) = f\nabla\cdot\mathbf{A} + \mathbf{A}\cdot\nabla f,$$

$$\nabla\times(f\mathbf{A}) = f\nabla\times\mathbf{A} + (\nabla f)\times\mathbf{A},$$

$$\nabla\times(f\mathbf{A}) = f\nabla\times\mathbf{A} + (\nabla f)\times\mathbf{A},$$

$$\nabla\cdot(\mathbf{A}\times\mathbf{B}) = (\nabla\times\mathbf{A})\mathbf{B} - (\nabla\times\mathbf{B})\mathbf{A},$$

$$\nabla\times(\mathbf{A}\times\mathbf{B}) = \mathbf{A}(\nabla\cdot\mathbf{B}) + (\mathbf{B}\cdot\nabla)\mathbf{A} - \mathbf{B}(\nabla\cdot\mathbf{A}) - (\mathbf{A}\cdot\nabla)\mathbf{B},$$

$$\mathbf{A}\times(\nabla\times\mathbf{B}) = (\nabla\mathbf{B})\cdot\mathbf{A} - (\mathbf{A}\cdot\nabla)\mathbf{B},$$

$$\nabla(\mathbf{A}\cdot\mathbf{B}) = \mathbf{A}\times(\nabla\times\mathbf{B}) + \mathbf{B}\times(\nabla\times\mathbf{A}) + (\mathbf{A}\cdot\nabla)\mathbf{B} + (\mathbf{B}\cdot\nabla)\mathbf{A},$$

$$\nabla\cdot\mathbf{A}\mathbf{B} = (\nabla\cdot\mathbf{A})\mathbf{B} + (\mathbf{A}\cdot\nabla)\mathbf{B}.$$

For a tensor $\mathbf{T}$ with Cartesian components T_{ij}, $\mathbf{T} = \sum_{ij} T_{ij}\mathbf{e}_i\mathbf{e}_j$:

$$(\nabla\cdot\mathbf{T})_j = \sum_i \frac{\partial T_{ij}}{\partial x_i},$$

$$\nabla\cdot(g\mathbf{T}) = (\nabla g)\cdot\mathbf{T} + g\nabla\cdot\mathbf{T},$$

$$\nabla^2 f = \nabla\cdot\nabla f,$$

$$\nabla^2\mathbf{A} = \nabla(\nabla\cdot\mathbf{A}) - \nabla\times(\nabla\times\mathbf{A}),$$

$$\nabla\times\nabla f = 0,$$

$$\nabla\cdot(\nabla\times\mathbf{A}) = 0.$$

If $\mathbf{r}$ is a position vector with magnitude $|\mathbf{r}| = r$,

$$\nabla\cdot\mathbf{r} = 3, \qquad \nabla\times\mathbf{r} = 0,$$

$$\nabla r = \frac{\mathbf{r}}{r}, \qquad \nabla\frac{1}{r} = -\frac{\mathbf{r}}{r^3},$$

$$\nabla^2\frac{1}{r} = -\nabla\cdot\left(\frac{\mathbf{r}}{r^3}\right) = 4\pi\delta(\mathbf{r}).$$

If V is a volume enclosed by a surface S and $d\mathbf{S} = \mathbf{n}\,dS$, where $\mathbf{n}$ is the unit normal pointing outward from the volume V,

$$\int_V \nabla(\cdots)\,dV = \int_S \mathbf{n}(\cdots)\,dS,$$

$$\int_V \nabla f\,dV = \int_S \mathbf{n}f\,dS,$$

$$\int_V \nabla\cdot\mathbf{A}\,dV = \int_S \mathbf{n}\cdot\mathbf{A}\,dS,$$

$$\int_V \nabla\cdot\mathbf{T}\,dV = \int_S \mathbf{n}\cdot\mathbf{T}\,dS,$$

$$\int_V \nabla\times\mathbf{A}\,dV = \int_S \mathbf{n}\times\mathbf{A}\,dS.$$

$$\int_V (f\nabla^2 g - g\nabla^2 f)\,dV = \int_S \mathbf{n}\cdot(f\nabla g - g\nabla f)\,dS,$$

$$\int_V [\mathbf{A}\cdot(\nabla\times(\nabla\times\mathbf{B})) - \mathbf{B}\cdot(\nabla\times(\nabla\times\mathbf{A}))]\,dV$$

$$= \int_S \mathbf{n}\cdot[\mathbf{B}\times(\nabla\times\mathbf{A}) - \mathbf{A}\times(\nabla\times\mathbf{B})]\,dS.$$

If S is an open surface bounded by the contour C of which the line element is $d\mathbf{s}$,

$$\int_S \mathbf{n} \times \nabla g \, dS = \oint_C g \, d\mathbf{s},$$

$$\int_S \mathbf{n} \cdot \nabla \times \mathbf{A} \, dS = \oint_C \mathbf{A} \cdot d\mathbf{s},$$

$$\int_S (\mathbf{n} \times \nabla) \times \mathbf{A} \, dS = -\oint_C \mathbf{A} \times d\mathbf{s},$$

$$\int_S \mathbf{n} \cdot (\nabla f \times \nabla g) \, dS = \oint_C f \nabla g \cdot d\mathbf{s} = -\oint_C g \nabla f \cdot d\mathbf{s}.$$

8.18   Orthogonal coordinate systems

In a general orthogonal coordinate system the three unit vectors are functions of position; therefore, when taking derivatives of vectors in such systems the derivatives of the unit vectors as well as the derivatives of the components must be considered. The element of distance can be written

$$d\mathbf{s} = h_1 \, dx_1 \mathbf{e}_1 + h_2 \, dx_2 \mathbf{e}_2 + h_3 \, dx_3 \mathbf{e}_3.$$

rectangular: $x_1 = x, \ x_2 = y, \ x_3 = z,$ $h_i = h_2 = h_3 = 1,$

cylindrical: $x_1 = \rho, \ x_2 = \phi, \ x_3 = z,$ $h_1 = h_3 = 1, \ h_2 = \rho,$

spherical: $x_1 = r, \ x_2 = \theta, \ x_3 = \phi,$ $h_1 = 1, \ h_2 = r, \ h_3 = r \sin\theta.$

8.18.1   Gradient of f, ∇f

$$\nabla f = \frac{\mathbf{e}_1}{h_1} \frac{\partial f}{\partial x_1} + \frac{\mathbf{e}_2}{h_2} \frac{\partial f}{\partial x_2} + \frac{\mathbf{e}_3}{h_3} \frac{\partial f}{\partial x_3}.$$

rectangular: $= \mathbf{e}_1 \dfrac{\partial f}{\partial x} + \mathbf{e}_2 \dfrac{\partial f}{\partial y} + \mathbf{e}_3 \dfrac{\partial f}{\partial z},$

cylindrical: $= \mathbf{e}_1 \dfrac{\partial f}{\partial \rho} + \dfrac{\mathbf{e}_2}{\rho} \dfrac{\partial f}{\partial \phi} + \mathbf{e}_3 \dfrac{\partial f}{\partial z},$

spherical: $= \mathbf{e}_1 \dfrac{\partial f}{\partial r} + \dfrac{\mathbf{e}_2}{r} \dfrac{\partial f}{\partial \theta} + \dfrac{\mathbf{e}_3}{r \sin\theta} \dfrac{\partial f}{\partial \phi}.$

8.18.2 Divergence of A, $\nabla \cdot A$

$$\nabla \cdot \mathbf{A} = \frac{1}{h_1 h_2 h_3} \sum_j \left[\frac{\partial}{\partial x_1}(A_1 h_2 h_3) + \frac{\partial}{\partial x_2}(h_1 A_2 h_3) + \frac{\partial}{\partial x_3}(h_1 h_2 A_3) \right],$$

rectangular: $= \dfrac{\partial A_1}{\partial x} + \dfrac{\partial A_2}{\partial y} + \dfrac{\partial A_3}{\partial z}.$

cylindrical: $= \dfrac{1}{\rho}\dfrac{\partial(\rho A_1)}{\partial \rho} + + \dfrac{1}{\rho}\dfrac{\partial A_2}{\partial \phi} + \dfrac{\partial A_3}{\partial z}.$

spherical: $= \dfrac{1}{r^2}\dfrac{\partial}{\partial r}(r^2 A_1) + \dfrac{1}{r \sin\theta}\left[\dfrac{\partial}{\partial \theta}(A_2 \sin\theta) + \dfrac{\partial A_3}{\partial \phi} \right].$

8.18.3 Curl of A, $\nabla \times A$

$$\nabla \times \mathbf{A} = \frac{1}{h_1 h_2 h_3} \begin{vmatrix} h_1 \mathbf{e}_1 & h_2 \mathbf{e}_2 & h_3 \mathbf{e}_3 \\ \dfrac{\partial}{\partial x} & \dfrac{\partial}{\partial y} & \dfrac{\partial}{\partial z} \\ h_1 A_1 & h_2 A_2 & h_3 A_3 \end{vmatrix}$$

$$= \frac{\mathbf{e}_1}{h_2 h_3}\left[\frac{\partial h_3 A_3}{\partial x_2} - \frac{\partial h_2 A_2}{\partial x_3} \right] + \frac{\mathbf{e}_2}{h_3 h_1}\left[\frac{\partial h_1 A_1}{\partial x_3} - \frac{\partial h_3 A_3}{\partial x_1} \right]$$

$$+ \frac{\mathbf{e}_3}{h_1 h_2}\left[\frac{\partial h_2 A_2}{\partial x_3} - \frac{\partial h_1 A_1}{\partial x_2} \right],$$

rectangular: $= \mathbf{e}_1 \left[\dfrac{\partial A_3}{\partial y} - \dfrac{\partial A_2}{\partial z} \right] + \mathbf{e}_2 \left[\dfrac{\partial A_1}{\partial z} - \dfrac{\partial A_3}{\partial x} \right]$

$$+ \mathbf{e}_3 \left[\frac{\partial A_2}{\partial x} - \frac{\partial A_1}{\partial y} \right],$$

cylindrical: $= \mathbf{e}_1 \left[\dfrac{1}{\rho}\dfrac{\partial A_3}{\partial \phi} - \dfrac{\partial A_2}{\partial z} \right] + \mathbf{e}_2 \left[\dfrac{\partial A_1}{\partial z} - \dfrac{\partial A_3}{\partial \rho} \right]$

$$+ \frac{\mathbf{e}_3}{\rho}\left[\frac{\partial(\rho A_2)}{\partial \rho} - \frac{\partial A_1}{\partial \phi} \right],$$

spherical: $= \dfrac{\mathbf{e}_1}{r \sin\theta}\left[\dfrac{\partial(A_3 \sin\theta)}{\partial \theta} - \dfrac{\partial A_2}{\partial \phi} \right]$

$$+ \frac{\mathbf{e}_2}{r \sin\theta}\left[\frac{\partial A_1}{\partial \phi} - \sin\theta \frac{\partial(r A_3)}{\partial r} \right] + \frac{\mathbf{e}_3}{r}\left[\frac{\partial(r A_2)}{\partial r} - \frac{\partial A_1}{\partial \theta} \right].$$

8.18.4   Projected derivative, components of $\mathbf{A} \cdot \nabla \mathbf{B}$

$$\mathbf{A} \cdot \nabla \mathbf{B} = \sum_j \mathbf{e}_j \left[\mathbf{A} \cdot \nabla B_j + \sum_i \left(A_j \frac{\partial h_j}{\partial x_i} - A_i \frac{\partial h_i}{\partial x_j} \right) \frac{B_i}{h_i h_j} \right],$$

rectangular: $\displaystyle = \mathbf{e}_1 \left[A_1 \frac{\partial B_1}{\partial x} + A_2 \frac{\partial B_1}{\partial y} + A_3 \frac{\partial B_1}{\partial z} \right]$

$$+ \mathbf{e}_2 \left[A_1 \frac{\partial B_2}{\partial x} + A_2 \frac{\partial B_2}{\partial y} + A_3 \frac{\partial B_2}{\partial z} \right]$$

$$+ \mathbf{e}_3 \left[A_1 \frac{\partial B_3}{\partial x} + A_2 \frac{\partial B_3}{\partial y} + A_3 \frac{\partial B_3}{\partial z} \right],$$

cylindrical: $\displaystyle = \mathbf{e}_1 \left[A_1 \frac{\partial B_1}{\partial \rho} + \frac{A_2}{\rho} \frac{\partial B_1}{\partial \phi} + A_3 \frac{\partial B_1}{\partial z} - \frac{A_2 B_2}{\rho} \right]$

$$+ \mathbf{e}_2 \left[A_1 \frac{\partial B_2}{\partial \rho} + \frac{A_2}{\rho} \frac{\partial B_2}{\partial \phi} + A_3 \frac{\partial B_2}{\partial z} + \frac{A_2 B_1}{\rho} \right]$$

$$+ \mathbf{e}_3 \left[A_1 \frac{\partial B_3}{\partial \rho} + \frac{A_2}{\rho} \frac{\partial B_2}{\partial \phi} + A_3 \frac{\partial B_3}{\partial z} \right],$$

spherical: $\displaystyle = \mathbf{e}_1 \left[A_1 \frac{\partial B_1}{\partial r} + \frac{A_2}{r} \frac{\partial B_1}{\partial \theta} + \frac{A_3}{r \sin \theta} \frac{\partial B_1}{\partial \phi} - \frac{A_2 B_2 + A_3 B_3}{r} \right]$

$$+ \mathbf{e}_2 \left[A_1 \frac{\partial B_2}{\partial r} + \frac{A_2}{r} \frac{\partial B_2}{\partial \theta} + \frac{A_3}{r \sin \theta} \frac{\partial B_2}{\partial \phi} + \frac{A_2 B_1 - A_3 B_3 \cot \theta}{r} \right]$$

$$+ \mathbf{e}_3 \left[A_1 \frac{\partial B_3}{\partial r} + \frac{A_2}{r} \frac{\partial B_3}{\partial \theta} + \frac{A_3}{r \sin \theta} \frac{\partial B_3}{\partial \phi} + \frac{A_3}{r} (B_1 + B_2 \cot \theta) \right].$$

8.18.5   Divergence of a tensor, $\nabla \mathbf{T}$

In an orthogonal coordinate system, a tensor $\mathbf{T}$ can be written in dyadic form $\mathbf{T} = \sum_{ij} T_{ij} \mathbf{e}_i \mathbf{e}_j$

$$\nabla \cdot \mathbf{T} = \sum_j \mathbf{e}_j \left[\nabla \cdot T_{\cdot j} + \sum_i \frac{1}{h_j h_i} \left(T_{ji} \frac{\partial h_j}{\partial x_i} - T_{ii} \frac{\partial h_i}{\partial x_j} \right) \right],$$

rectangular: $= \mathbf{e}_1 \left[\dfrac{\partial T_{11}}{\partial x} + \dfrac{\partial T_{21}}{\partial y} + \dfrac{\partial T_{31}}{\partial z} \right]$

$+ \mathbf{e}_2 \left[\dfrac{\partial T_{12}}{\partial x} + \dfrac{\partial T_{22}}{\partial y} + \dfrac{\partial T_{32}}{\partial z} \right]$

$+ \mathbf{e}_3 \left[\dfrac{\partial T_{13}}{\partial x} + \dfrac{\partial T_{23}}{\partial y} + \dfrac{\partial T_{33}}{\partial z} \right].$

cylindrical: $= \mathbf{e}_1 \left[\dfrac{\partial T_{11}}{\partial \rho} + + \dfrac{1}{\rho} \dfrac{\partial T_{21}}{\partial \phi} + \dfrac{\partial T_{31}}{\partial z} + \dfrac{T_{11} - T_{22}}{\rho} \right]$

$+ \mathbf{e}_2 \left[\dfrac{\partial T_{12}}{\partial \rho} + \dfrac{1}{\rho} \dfrac{\partial T_{22}}{\partial \phi} + \dfrac{\partial T_{32}}{\partial z} + \dfrac{T_{12} + T_{21}}{\rho} \right]$

$+ \mathbf{e}_3 \left[\dfrac{\partial T_{13}}{\partial \rho} + \dfrac{T_{13}}{\rho} + \dfrac{1}{\rho} \dfrac{\partial T_{23}}{\partial \phi} + \dfrac{\partial T_{33}}{\partial z} \right].$

spherical: $= \mathbf{e}_1 \left[\dfrac{\partial T_{11}}{\partial r} + \dfrac{1}{r \sin \theta} \left(\dfrac{\partial}{\partial \theta} (\sin \theta T_{21}) + \dfrac{\partial T_{31}}{\partial \phi} \right) \right.$

$\left. + \dfrac{2T_{11} - T_{22} - T_{33}}{r} \right]$

$+ \mathbf{e}_2 \left[\dfrac{\partial T_{12}}{\partial r} + \dfrac{1}{r \sin \theta} \left(\dfrac{\partial}{\partial \theta} (\sin \theta T_{22}) + \dfrac{\partial T_{32}}{\partial \phi} \right) \right.$

$\left. + \dfrac{2T_{12} + T_{21} - T_{33} \cot \theta}{r} \right]$

$+ \mathbf{e}_3 \left[\dfrac{\partial T_{13}}{\partial r} + \dfrac{1}{r \sin \theta} \left(\dfrac{\partial}{\partial \theta} (\sin \theta T_{23}) + \dfrac{\partial T_{33}}{\partial \phi} \right) \right.$

$\left. + \dfrac{2T_{13} + T_{31} + T_{32} \cot \theta}{r} \right].$

8.18.6 Scalar Laplacian, $\nabla^2 f$

$\nabla^2 f = \operatorname{div} \operatorname{grad} f$

$= \dfrac{1}{h_1 h_2 h_3} \left[\dfrac{\partial}{\partial x_1} \left(\dfrac{h_2 h_3}{h_1} \dfrac{\partial f}{\partial x_1} \right) + \dfrac{\partial}{\partial x_2} \left(\dfrac{h_3 h_1}{h_2} \dfrac{\partial f}{\partial x_2} \right) + \dfrac{\partial}{\partial x_3} \left(\dfrac{h_1 h_2}{h_3} \dfrac{\partial f}{\partial x_3} \right) \right].$

rectangular:
$$= \frac{\partial^2 f}{\partial x^2} + \frac{\partial^2 f}{\partial y^2} + \frac{\partial^2 f}{\partial z^2}.$$

cylindrical:
$$= \frac{1}{\rho} \frac{\partial}{\partial \rho} \left(\rho \frac{\partial f}{\partial \rho} \right) + \frac{1}{\rho^2} \frac{\partial^2 f}{\partial \phi^2} + \frac{\partial^2 f}{\partial z^2}.$$

spherical:
$$= \frac{1}{r^2} \frac{\partial}{\partial r} \left(r^2 \frac{\partial f}{\partial r} \right) + \frac{1}{r^2 \sin^2 \theta} \left[\sin \theta \frac{\partial}{\partial \theta} \left(\sin \theta \frac{\partial f}{\partial \theta} \right) + \frac{\partial^2 f}{\partial \phi^2} \right].$$

8.18.7 Vector Laplacian, $\nabla^2 \mathbf{A}$

$$\nabla^2 \mathbf{A} = \nabla(\nabla \cdot \mathbf{A}) - \nabla \times (\nabla \times \mathbf{A}).$$

rectangular:
$$= \mathbf{e}_1 \nabla^2 A_1 + \mathbf{e}_2 \nabla^2 A_2 + \mathbf{e}_3 \nabla^2 A_3$$

$$= \mathbf{e}_1 \left[\frac{\partial^2 A_1}{\partial x^2} + \frac{\partial^2 A_1}{\partial y^2} + \frac{\partial^2 A_1}{\partial z^2} \right]$$

$$+ \mathbf{e}_2 \left[\frac{\partial^2 A_2}{\partial x^2} + \frac{\partial^2 A_2}{\partial y^2} + \frac{\partial A_2}{\partial z^2} \right] + \mathbf{e}_3 \left[\frac{\partial^2 A_3}{\partial x^2} + \frac{\partial^2 A_3}{\partial y^2} + \frac{\partial^2 A_3}{\partial z^2} \right].$$

cylindrical:
$$= \mathbf{e}_1 \left[\nabla^2 A_1 - \frac{1}{\rho^2} \left(2 \frac{\partial A_2}{\partial \phi} + A_1 \right) \right]$$

$$+ \mathbf{e}_2 \left[\nabla^2 A_2 + \frac{1}{\rho^2} \left(2 \frac{\partial A_1}{\partial \phi} - A_2 \right) \right] + \mathbf{e}_3 \left[\nabla^2 A_3 \right].$$

spherical:
$$= \mathbf{e}_1 \left[\nabla^2 A_1 - \frac{2}{r^2} \left(A_1 + \frac{\partial A_2}{\partial \theta} + \cot \theta A_2 + \csc \theta \frac{\partial A_3}{\partial \phi} \right) \right]$$

$$+ \mathbf{e}_2 \left[\nabla^2 A_2 + \frac{2}{r^2} \frac{\partial A_1}{\partial \theta} - \frac{1}{r^2 \sin^2 \theta} \left(2 \cos \theta \frac{\partial A_3}{\partial \phi} + A_2 \right) \right]$$

$$+ \mathbf{e}_3 \left[\nabla^2 A_3 + \frac{1}{r^2 \sin^2 \theta} \left(2 \sin \theta \frac{\partial A_1}{\partial \phi} + 2 \cos \theta \frac{\partial A_2}{\partial \phi} - A_3 \right) \right].$$

8.19   General curvilinear spaces

The Einstein convention is used: the explicit summation sign is suppressed and an index appearing as both a subscript and a superscript is a dummy value to be summed over its range.

8.19.1   Metric tensor

The metric of an n-space is given by

$$ds^2 = g_{\mu\nu}\, dx^\mu dx^\nu,$$

where dx^μ are generalized coordinates (a set of numbers locating a point in the space) and $g_{\mu\nu}$ is the metric tensor:

$$g^{\mu\sigma} g_{\sigma\nu} = g^\mu_\nu = \delta^\mu_\nu.$$

(The tensor $g^{\mu\nu}$ is the inverse of the tensor $g_{\mu\nu}$.)

Raising and Lowering Indices: The dual vector is defined by $A_\mu = g_{\mu\nu} A^\nu$; in terms of the components of the displacement vector and its dual, the differential distance in the space is expressed as

$$ds^2 = dx_\mu dx^\mu.$$

For any tensor $T^{\cdot\mu\cdot\nu\cdots}_{\alpha\cdot\beta\cdots}$,

$$g_{\sigma\mu} T^{\cdot\mu\cdot\nu\cdots}_{\alpha\cdot\beta\cdots} = T^{\cdots\nu\cdots}_{\alpha\sigma\beta\cdots}, \qquad g^{\gamma\alpha} T^{\cdot\mu\cdot\nu\cdots}_{\alpha\cdot\beta\cdots} = T^{\gamma\mu\cdot\nu\cdots}_{\cdot\cdot\beta\cdots}.$$

8.19.2   Characterization of intrinsic curvature

Christoffel Symbols: $[\mu\nu; \sigma] = [\nu\mu; \sigma]$,

$$[\mu\nu; \sigma] = \frac{1}{2}\left(\frac{\partial g_{\sigma\nu}}{\partial x^\mu} + \frac{\partial g_{\mu\sigma}}{\partial x^\nu} - \frac{\partial g_{\mu\nu}}{\partial x^\sigma}\right),$$

$$\Gamma^\sigma_{\mu\nu} = [\mu\nu; \lambda] g^{\lambda\sigma},$$

$$\Gamma^a_{a\mu} = \frac{\partial\sqrt{|g|}}{\partial x^\mu}, \qquad g = \det(g_{\mu\nu}).$$

Covariant Differentiation: $\dfrac{dA^{a\cdots}_b}{dt} \equiv A^{a\cdots}_{b\cdots;c}\,\dfrac{dx^c}{dt}$,

$$A^a_{;c} = \frac{\partial A^a}{\partial x^c} + \Gamma^a_{cd} A^d,$$

$$A_{b;c} = \frac{\partial A_b}{\partial x^c} - \Gamma^d_{bc} A_d,$$

$$T^{ab\cdots}_{cd\cdots;s} = \frac{\partial T^{ab\cdots}_{cd\cdots}}{\partial x^s} + \Gamma^a_{rs} T^{rb\cdots}_{cd\cdots} + \Gamma^b_{rs} T^{ar\cdots}_{cd\cdots} + \cdots$$
$$- \Gamma^r_{cs} T^{ab\cdots}_{rd\beta\cdots} - \Gamma^r_{ds} T^{ab\cdots}_{cr\cdots} - \cdots,$$

$$g_{ab;s} = \frac{\partial g_{ab}}{\partial x^s} - \Gamma^r_{as} g_{rb} - \Gamma^r_{sb} g_{ar} \equiv 0.$$

This last condition is the defining relation for Γ^s_{ab}.

Riemann-Christoffel Tensor: Covariant differentiation is non-commutative if the space is not flat.

$$A_{c;ba} - A_{c;ab} = R^d{}_{cba} A_d,$$

$$R^d{}_{cba} = \frac{\partial \Gamma^d_{ac}}{\partial x^b} - \frac{\partial \Gamma^d_{bc}}{\partial x^a} + \Gamma^e_{ac}\Gamma^d_{eb} - \Gamma^e_{bc}\Gamma^d_{ea},$$

$$R^d{}_{abc} + R^d{}_{bca} + R^d{}_{cab} = 0,$$

$$R_{dcba} = g_{de}R^e{}_{cba} = -R_{cdba} = -R_{dcab} = R_{abcd}.$$

As a result of these symmetries, only $n^2(n^2-1)/12$ of the n^4 components in n-space are independent.

Ricci Tensor:

$$R_{ca} = R^d{}_{cda} = \frac{1}{\sqrt{|g|}}\frac{\partial}{\partial x^d}\left(\sqrt{|g|}\Gamma^d_{ac}\right) - \frac{\partial^2 \ln\sqrt{|g|}}{\partial x^a \partial x^c} - \Gamma^d_{ae}\Gamma^e_{dc} = R_{ac}.$$

The choice of contraction is essentially unique since $R^d{}_{dbc} = 0$ and $R^d{}_{cbd} = -R^d{}_{cdb}$.

The contraction of the Ricci tensor: $R = R^a_a = g^{ac}R_{ac}$ is the mean curvature (the *Gaussian* curvature of the surface for $n = 2$).

8.20   Fourier series and Fourier transforms

A function $f(x)$ in the interval $-a/2 \le x \le a/2$ may be expanded in a trigonometric series

$$f(x) = \frac{1}{2}A_0 + \sum_{m=1}^{\infty}\left[A_m \cos\frac{2\pi mx}{a} + B_m \sin\frac{2\pi mx}{a}\right],$$

where

$$A_m = \frac{2}{a}\int_{-a/2}^{a/2} f(x)\cos\frac{2\pi mx}{a}\,dx,$$

$$B_m = \frac{2}{a}\int_{-a/2}^{a/2} f(x)\sin\frac{2\pi mx}{a}\,dx.$$

The functions

$$\sqrt{\frac{2}{a}}\sin\frac{2\pi mx}{a} \qquad \text{and} \qquad \sqrt{\frac{2}{a}}\cos\frac{2\pi mx}{a}$$

form an orthogonal set.

Table 8.1 Fourier transform pairs.   Additional transforms may be
obtained by interchanging $f(x)$ and $F(k)$, by differentiating these expressions, by applying the convolution relationship, and by using other general
properties.

$f(x)$	$F(k)$	Conditions
e^{iax}	$\sqrt{2\pi}\delta(k+a)$	a real
$\exp(-a^2x^2/2)$	$\dfrac{1}{a}\exp(-k^2/2a^2)$	$a>0$
1	$\sqrt{2\pi}\delta(k)$	
$\dfrac{1}{x}$	$\sqrt{2\pi}\,i\,\mathrm{sgn}\,k$	
$\dfrac{1}{|x|}$	$\dfrac{1}{|k|}$	
$|x|^{-a}$	$\sqrt{2\pi}\Gamma(1-a)\sin\left(\dfrac{\pi a}{2}\right)$	$0<a<1$
$e^{-a|x|}$	$\dfrac{1}{\sqrt{\pi}}\dfrac{a^2}{\pi(a^2+x^2)}$	$a>0$
$\sin\dfrac{a^2x^2}{2}$	$\dfrac{1}{\sqrt{2}\,a}\left(\cos\dfrac{k^2}{2a^2}-\sin\dfrac{k^2}{a^2}\right)$	
$\cos\dfrac{a^2x^2}{2}$	$\dfrac{1}{\sqrt{2}\,a}\left(\cos\dfrac{k^2}{2a^2}+\sin\dfrac{k^2}{a^2}\right)$	
$\dfrac{\sin(ax)}{x}$	$\begin{cases} \sqrt{\dfrac{\pi}{2}} & \text{for } |k|<a \\ 0 & \text{for } |k|>a \end{cases}$	$a>0$

When the interval becomes infinite, the expansion becomes the *Fourier
integral*

$$f(x) = \frac{1}{\sqrt{2\pi}}\int_{-\infty}^{\infty} F(k)e^{-ikx}\,dk,$$

where

$$F(k) = \mathcal{F}[f(x);k] = \frac{1}{\sqrt{2\pi}}\int_{-\infty}^{\infty} f(x)e^{ikx}\,dx.$$

is the *Fourier transform* of $f(x)$.

If $F(k)$ is the Fourier transform of $f(x)$, then $f^*(k)$ is the Fourier transform of $F^*(x)$ (the asterisk indicating complex conjugate). For arbitrary
fixed constants a and b,

$$\mathcal{F}[f(ax);k] = \frac{1}{a}F\left(\frac{k}{a}\right), \qquad \mathcal{F}[af(x)+bg(x);k] = aF(k)+bG(k),$$

Table 8.2 Laplace transforms. Additional transforms may be obtained by making use of the general properties and combinations of them, by the replacement $a \to ia$, etc.

$F(s)$	$f(t)$	
$1/s$	$U(t)$	
$\sqrt{\dfrac{1}{s}}$	$\dfrac{1}{\sqrt{\pi t}}$	
$\dfrac{1}{s^k}$	$\dfrac{t^{k-1}}{\Gamma(k)}$	$k > 0$
$\dfrac{1}{s+a}$	e^{-at}	
$\dfrac{1}{(s+a)(s+b)}$	$\dfrac{e^{-at} - e^{-bt}}{b-a}$	$a \neq b$
$\dfrac{s}{(s+a)(s+b)}$	$\dfrac{ae^{-at} - be^{-bt}}{a-b}$	$a \neq b$
$\dfrac{a}{s^2 + a^2}$	$\sin at$	
$\dfrac{s}{s^2 + a^2}$	$\cos at$	
$\dfrac{1}{s^3 + a^3}$	$\dfrac{1}{3a^2}\left[e^{-at} - e^{\frac{1}{2}at}\left(\cos\dfrac{\sqrt{3}\,at}{2} - \sqrt{3}\sin\dfrac{\sqrt{3}\,at}{2} \right)\right]$	

$$\mathcal{F}[f(x+s); k] = e^{-iks} F(k),$$

If $\displaystyle\lim_{|x|\to\infty} f'(x) = 0,$ $\quad \mathcal{F}[f'(x); k] = ikF(k).$

The *convolution* of two functions $f(x)$ and $g(x)$ is defined as

$$(f \circ g)(x) \equiv \int_{-\infty}^{\infty} f(y)g(x-y)\,dy$$

with the properties

$$(f \circ g)(x) = (g \circ f)(x), \qquad \big((f \circ g) \circ h\big)(x) = \big(f \circ (g \circ h)\big)(x),$$

$$\mathcal{F}[(f \circ g)(x); k] = F(k)g(k), \qquad \mathcal{F}[f(x)g(x)] = (F \circ G)(k).$$

8.21 Laplace transforms

If $f(t)$ is a function of the real variable t, $t > 0$; the *Laplace Transform* of $f(t)$ is

$$\mathcal{L}\{f(t)\} = F(s) = \int_0^\infty f(t)e^{-st}\,dt,$$

where s is a complex variable. If the integral is convergent for some real value of s, $s = s_0$, then it converges for all s with $\Re s > s_0$, and $F(s)$ is a single valued analytic function of s in the half-plane $\Re s > s_0$.

8.21.1 General properties

Inversion: $f(t) = \dfrac{1}{2\pi i} \displaystyle\int_{c-i\infty}^{c+i\infty} e^{ts} F(s)\, ds.$

Linearity: $\mathcal{L}\{af(t) + bg(t)\} = aF(s) + bG(s).$

Differentiation: $\mathcal{L}\{f'(t)\} = sF(s) - f(+0),$

$$\mathcal{L}\{f^{(n)}(t)\} = s^n F(s) - s^{n-1} f(+0) - s^{n-2} f'(+0) - \cdots - f^{(n-1)}(+0),$$

$$n \quad \text{integer, } n > 0,$$

$$\mathcal{L}\{t^n f(t)\} = (-)^n F^{(n)}(s).$$

Integration: $\mathcal{L}\left\{ \displaystyle\int_0^t f(t')\, dt' \right\} = \dfrac{1}{s} F(s),$

$$\mathcal{L}\left\{ \frac{1}{t} f(t) \right\} = \int_s^\infty F(x)\, dx.$$

Convolution: $\mathcal{L}\{(f_1 \circ f_2)(t)\} = \mathcal{L}\{\int_0^\infty f_1(t-t') f_2(t')\, dt'\} = F_1(s) F_2(s).$

Translation and Scaling:

$$\mathcal{L}\{e^{at} f(t)\} = F(s-a),$$

$$\mathcal{L}\{f(t/c)\} = cF(cs), \qquad\qquad c > 0,$$

$$\mathcal{L}\{e^{at} f(bt)\} = F\big((s-a)/b\big), \qquad\quad b > 0,$$

$$\mathcal{L}\{f(t-b) U(t-b)\} = e^{-bs} F(s), \qquad\quad b > 0,$$

where

$$U(t) = \begin{cases} 0 & t < 0 \\ \frac{1}{2} & t = 0 \\ 1 & t > 0 \end{cases}.$$

Periodic functions: If $f(t)$ is periodic with period T: $f(t+T) = f(t)$ for all t,

$$\mathcal{L}\{f(t)\} = \frac{\int_0^T e^{-st} f(t)\, dt}{1 - e^{-sT}}.$$

8.22   Bessel functions

8.22.1   Differential equation

$$z^2 \frac{d^2 w}{dz^2} + z \frac{dw}{dz} + (z^2 - \nu^2)w = z \frac{d}{dz}\left(z \frac{dw}{dz}\right) + (z^2 - \nu^2)w = 0.$$

The argument z and the order ν are complex; w is regular throughout the z-plane cut along the negative real axis with a branch point at the origin. Solutions of this equation are the Bessel functions of the first kind $J_\nu(z)$, the Weber functions (Bessel functions of the second kind) $Y_\nu(z)$, and the Hankel functions (Bessel functions of the third kind) $H_\nu^{(1)}(z)$, $H_\nu^{(2)}(z)$.

$J_\nu(z)$ and $J_{-\nu}(z)$ are linearly independent solutions when ν is not an integer; for $\nu = \pm n$ (n is a positive integer or 0), $J_{-n}(z) = (-1)^n J_n(z)$, there is no branch point at $z = 0$ and $J_n(z)$ is an entire function of z.

$$Y_\nu(z) = N_\nu(z) = \frac{J_\nu(z)\cos(\nu z) - J_\nu(z)}{\sin(\nu z)},$$

$$Y_n(z) = \lim_{\epsilon \to 0} Y_{n+\epsilon}(z),$$

$$H_\nu^{(1)}(z) = J_\nu(z) + iY_\nu(z) = i\frac{e^{-i\pi\nu} J_\nu(z) - J_{-\nu}(z)}{\sin(\nu z)},$$

$$H_\nu^{(2)}(z) = J_\nu(z) - iY_\nu(z) = i\frac{J_{-\nu}(z) - e^{i\pi\nu} J_\nu(z)}{\sin(\nu z)},$$

$$H_{-\nu}^{(1)}(z) = e^{i\pi\nu} H^{(1)}(z), \qquad H_{-\nu}^{(2)}(z) = e^{-i\pi\nu} H^{(2)}(z).$$

The modified Bessel functions $I_{\pm\nu}(z)$ and $K_\nu(z)$ are solutions of the differential equation

$$z^2 \frac{d^2 v}{dz^2} + z \frac{dv}{dz} - (z^2 + \nu^2)v = z \frac{d}{dz}\left(z \frac{dv}{dz}\right) - (z^2 + \nu^2)v = 0,$$

$$K_\nu(z) = \frac{\pi}{2} \frac{I_{-\nu}(z) - I_\nu(z)}{\sin(\nu\pi)},$$

$$I_{-n}(z) = I_n(z), \qquad K_{-\nu}(z) = K_\nu(z),$$

$$I_\nu(z) = \begin{cases} e^{-i\nu\pi/2} J_\nu(e^{i\pi/2}z) & -\pi < \arg z \le \pi/2 \\ e^{3i\nu\pi/2} J_\nu(e^{-3i\pi/2}z) & \pi/2 < \arg z \le \pi \end{cases},$$

$$K_\nu(z) = \begin{cases} \frac{i\pi}{2} e^{i\nu\pi/2} H_\nu^{(1)}(e^{i\pi/2}z) & -\pi < \arg z \le \pi/2 \\ -\frac{i\pi}{2} e^{-i\nu\pi/2} H_\nu^{(1)}(e^{-i\pi/2}z) & \pi/2 < \arg z \le \pi \end{cases}.$$

For $z \to 0$, $\quad J_\nu(z) \sim \left(\frac{1}{2}z\right)^\nu / \Gamma(\nu + 1)$,

$$Y_\nu(z) \sim -iH_\nu^{(1)}(z) \sim iH_\nu^{(2)} \sim \Gamma(\nu)(2/z)^\nu / \pi, \quad \Re\nu > 0,$$

$$Y_0(z) \sim -iH_0^{(1)}(z) \sim iH_0^{(2)} \sim (2/\pi) \ln z;$$

for $x \to \infty$, $\quad J_\nu(x) \sim \sqrt{2/\pi x} \, \cos[x - (2\nu + 1)\pi/4]$,

$$Y_\nu(x) \sim \sqrt{2/\pi x} \, \sin[x - (2\nu + 1)\pi/4],$$

$$H_\nu^{(1)}(x) \sim \sqrt{2/\pi x} \, e^{i[x - (2\nu+1)/4]},$$

$$H_\nu^{(2)}(x) \sim \sqrt{2/\pi x} \, e^{-i[x - (2\nu+1)/4]}.$$

8.22.2 Series expansion

$$J_\nu(z) = \sum_{k=0}^{\infty} (-)^k \frac{\left(\frac{1}{2}z\right)^{2k+\nu}}{\Gamma(\nu + k + 1)k!},$$

$$I_\nu(z) = \sum_{k=0}^{\infty} \frac{\left(\frac{1}{2}z\right)^{2k+\nu}}{\Gamma(\nu + k + 1)k!},$$

$$Y_n(z) = -\frac{\left(\frac{1}{2}z\right)^{-n}}{\pi} \sum_{k=0}^{n-1} \frac{(n - k - 1)!}{k!} (\tfrac{1}{2}z)^{2k}$$

$$+ \frac{2}{\pi}\left[J_n(z) \ln\left(\frac{1}{2}z\right) - \sum_{k=1}^{\infty} C_k^n \frac{\left(\frac{1}{2}z\right)^{2k+n}}{(n + k)!k!} \right],$$

where

$$C_k^n = (-)^k \left[\psi(k + 1) + \frac{1}{2} \sum_{p=1}^{n} \frac{1}{k + p} \right].$$

$$\psi(1) = -\gamma, \quad \text{and} \quad \psi(k + 1) = \sum_{p=1}^{k} \frac{1}{p} - \gamma \quad \text{for} \quad k > 0.$$

$$J_0(z) = 1 - \left(\frac{1}{2}z\right)^2 + \frac{\left(\frac{1}{2}z\right)^4}{2!^2} - \frac{\left(\frac{1}{2}z\right)^6}{3!^2} + \frac{\left(\frac{1}{2}z\right)^8}{4!^2} - \cdots,$$

$$Y_0(z) = \frac{2}{\pi}\left[\ln\left(\frac{1}{2}z\right) + \gamma \right] J_0(z) + \left(\frac{1}{2}z\right)^2 - \left(1 + \frac{1}{2}\right)\frac{\left(\frac{1}{2}z\right)^4}{2!^2}$$

$$+ \left(1 + \frac{1}{2} + \frac{1}{3}\right)\frac{\left(\frac{1}{2}z\right)^6}{3!^2} + \left(1 + \frac{1}{2} + \frac{1}{3} + \frac{1}{4}\right)\frac{\left(\frac{1}{2}z\right)^8}{4!^2} - \cdots.$$

8.22.3 Integral representations

$$J_0(z) = \frac{1}{\pi} \int_0^\pi \cos(z \sin \theta) \, d\theta,$$

$$Y_0(z) = \frac{4}{\pi^2} \int_0^{\pi/2} \cos(z \cos \theta)[\gamma + \ln(2z \sin^2 \theta)] \, d\theta,$$

$$I_0(z) = \frac{1}{\pi} \int_0^\pi \cosh(z \sin \theta) \, d\theta,$$

$$K_0(z) = -\frac{1}{\pi} \int_0^\pi \cosh(z \cos \theta)[\gamma + \ln(2z \sin^2 \theta)] \, d\theta,$$

$$J_n(z) = \frac{z^n}{2\pi(2n-1)!!} \int_0^\pi \cos(z \sin \theta) \cos^{2n} \theta \, d\theta$$

$$= \frac{z^n}{\pi(2n-1)!!} \int_0^1 (1-t^2)^{n-1/2} \cos(zt) \, d\theta$$

$$= \frac{1}{\pi} \int_0^\pi \cos(z \sin \theta - n\theta) \, d\theta (|\arg z| < \frac{1}{2}\pi),$$

$$Y_n(z) = \frac{1}{\pi} \int_0^\pi \cos(z \sin \theta - n\theta) \, d\theta (|\arg z| < \frac{1}{2}\pi),$$

$$K_n(z) = \int_0^\infty e^{-z \cosh t} \cosh(nt) \, dt (|\arg z| < \frac{1}{2}\pi).$$

8.22.4 Recursion relations

Let $C_\nu(z)$ be any linear combination of J_ν, Y_ν, $H_\nu^{(1)}$, and $H_\nu^{(2)}$ with coefficients independent of z and ν:

$$C_{\nu-1}(z) + C_{\nu+1}(z) = \frac{2\nu}{z} C_\nu(z),$$

$$C_{\nu-1}(z) - C_{\nu+1}(z) = \frac{2\nu}{z} C_\nu'(z),$$

$$C_{\nu-1}(z) - \frac{\nu}{z} C_\nu(z) = C_\nu'(z),$$

$$\frac{\nu}{z} C_\nu(z) - C_{\nu+1}(z) = C_\nu'(z),$$

$$C_0'(z) = -C_1(z).$$

8.22.5 Generating functions and series

$$e^{z(t-1/t)/2} = \sum_{k=-\infty}^{\infty} t^k J_k(z), \qquad t \neq 0,$$

$$\cos(z \sin\theta) = J_0(z) + 2\sum_{k=1}^{\infty} J_{2k}(z)\cos(2k\theta),$$

$$\sin(z \sin\theta) = 2\sum_{k=0}^{\infty} J_{2k+1}(z)\sin[(2k+1)\theta],$$

$$e^{z\cos\theta} = I_0(z) + 2\sum_{k=1}^{\infty} I_k(z)\cos(k\theta),$$

$$e^{z\sin\theta} = I_0(z) + 2\sum_{k=1}^{\infty} (-)^k \big[I_{2k}(z)\cos(2k\theta)$$
$$- I_{2k-1}(z)\sin(2k-1)\theta\big].$$

8.22.6 Addition theorems

$$J_n(u \pm v) = \sum_{k=-\infty}^{\infty} (\pm 1)^k J_{n-k}(u) J_k(v).$$

For $w^2 = u^2 - 2uv\cos\theta + v^2$,

$$w\cos\phi = u - v\cos\alpha, \ \ w\sin\phi = v\sin\alpha,$$

$$J_n(w)e^{in\phi} = \sum_{k=-\infty}^{\infty} J_{n+k}(u) J_k(v) e^{ik\alpha}.$$

8.23 Spherical harmonics

8.23.1 Legendre polynomials, $P_n(x)$

Definition: $$\frac{1}{(1 - 2xt + t^2)^{1/2}} = \sum_{n=0}^{\infty} P_n(x)t^n.$$

$$P_n = \frac{1}{2^n n!}\frac{d^n}{dx^n}(x^2 - 1)^n,$$

$$(1 - x^2)P_n'' - 2xP_n' + n(n+1)P_n = 0,$$

$$P_{n+1}(x) = \frac{1}{n+1}[(2n+1)P_n(x) - nP_{n-1}(x)],$$

$$P_n(1) = 1, \qquad P_n(-1) = (-1)^n,$$

$$P_0(x) = 1, \qquad\qquad P_4(x) = \tfrac{1}{8}(35x^4 - 30x^2 + 3),$$

$$P_1(x) = x, \qquad\qquad P_5(x) = \tfrac{1}{8}(63x^5 - 70x^4 + 15x),$$

$$P_2(x) = \tfrac{1}{2}(3x^2 - 1), \qquad P_6(x) = \tfrac{1}{16}(231x^6 - 315x^4 + 105x^2 - 5),$$

$$P_3(x) = \tfrac{1}{2}(5x^3 - 3x),$$

$$\int_{-1}^{1} P_n(x) P_{n'}(x)\, dx = \frac{2\delta_{n,n'}}{n + n' + 1},$$

$$\int_{-1}^{1} x P_n(x) P_{n'}(x)\, dx = \begin{cases} \dfrac{n + n' + 1}{(n + n')(n + n' + 2)} & \text{if } |n - n'| = 1 \\ 0 & \text{otherwise} \end{cases}.$$

Potential at **r** *Due to a Unit Source at* **r′**:

$$\frac{1}{|\mathbf{r} - \mathbf{r'}|} = \frac{1}{r_>} \sum_{n=0}^{\infty} \left(\frac{r_<}{r_>}\right)^n P_n(\gamma), \qquad \gamma = \frac{\mathbf{r}\cdot\mathbf{r'}}{rr'} = \cos\theta,$$

where $r_<$ is the smaller and $r_>$ is the larger of the two distances, $|\mathbf{r}|$ and $|\mathbf{r'}|$, and γ is the cosine of the angle between the two position vectors.

8.23.2 Tesseral harmonics

Definition: $\quad Y_{l,m} = N_{l,m} e^{im\phi} \sin^{|m|}\theta \dfrac{d^{|m|} P_l(\cos\theta)}{d(\cos\theta)^{|m|}},$

where, for $m \geq 0$,

$$N_{l,m} = (-1)^m \sqrt{\frac{(2l + 1)}{4\pi} \frac{(l - m)!}{(l + m)!}} \qquad N_{l,-m} = (-1)^m N_{l,m}.$$

Normalization:

$$\iint Y_{l,m}^*(\theta, \phi) Y_{l',m'}(\theta, \phi)\, d\Omega = \delta_{l,l'}\delta_{m,m'},$$

where the asterisk denotes the complex conjugate.

Differential Equation:

$$\nabla^2[r^l Y_{l,m}(\theta, \phi)] = 0, \qquad \nabla^2[r^{-(l+1)} Y_{l,m}(\theta, \phi)] = 0, \qquad r \neq 0,$$

$$\nabla^2 = \frac{\partial^2}{\partial r^2} + \frac{2}{r}\frac{\partial}{\partial r} + \Lambda, \qquad \Lambda = \frac{1}{\sin\theta}\frac{\partial}{\partial\theta}\sin\theta\frac{\partial}{\partial\theta} + \frac{1}{\sin^2\theta}\frac{\partial^2}{\partial\phi^2}.$$

Completeness Relation:

$$\sum_{l=0}^{\infty} \sum_{m=-l}^{l} Y_{l,m}^{*}(\theta, \phi) Y_{l,m}(\theta', \phi') = \delta(\theta - \theta')\delta(\phi - \phi').$$

Specific Expressions:

$$Y_{0,0} = \sqrt{\frac{1}{4\pi}},$$

$$Y_{1,0} = \sqrt{\frac{3}{4\pi}} \cos\theta, \qquad\qquad Y_{3,0} = \frac{1}{2}\sqrt{\frac{7}{4\pi}}(5\cos^2\theta - 3)\cos\theta,$$

$$Y_{1,\pm1} = m_{\mathrm{p}}\sqrt{\frac{3}{8\pi}} \sin\theta e^{\pm i\phi}, \qquad Y_{3,\pm1} = m_{\mathrm{p}}\frac{1}{4}\sqrt{\frac{21}{4\pi}} \sin\theta(5\cos^2\theta - 1)e^{\pm i\phi},$$

$$Y_{2,0} = \frac{1}{2}\sqrt{\frac{5}{4\pi}}(3\cos^2\theta - 1), \qquad Y_{3,\pm2} = \frac{1}{4}\sqrt{\frac{105}{2\pi}} \sin^2\theta \cos\theta e^{\pm 2i\phi},$$

$$Y_{2,\pm1} = m_{\mathrm{p}}\sqrt{\frac{15}{8\pi}} \sin\theta \cos\theta e^{\pm i\phi}, \qquad Y_{3,\pm3} = m_{\mathrm{p}}\frac{1}{4}\sqrt{\frac{35}{4\pi}} \sin^3\theta e^{\pm 3i\phi},$$

$$Y_{2,\pm2} = \frac{1}{4}\sqrt{\frac{15}{2\pi}} \sin^2\theta e^{\pm 2i\phi}.$$

Boundary Value Problem: The general solution for a boundary value problem in spherical coordinates can be written as

$$\Phi(r, \theta, \phi) = \sum_{l=0}^{\infty} \sum_{m=-l}^{l} \left(A_{lm}r^l + \frac{B_{lm}}{r^{l+1}} \right) Y_{l,m}(\theta, \phi).$$

If the origin is not excluded and there are no singularities or sources within the sphere, $B_{lm} = 0$. If the potential on the spherical surface $r = a$ is given, $\Phi(a, \theta, \phi)$,

$$A_{lm} = \frac{F_{lm}}{a^l},$$

where

$$F_{lm} = \int\int Y_{l,m}^{*}(\theta, \phi)\Phi(a, \theta, \phi)\, d\Omega.$$

When the boundaries of the problem are two concentric spherical surfaces and the potential on the inner surface is $\Phi(b, \theta, \phi) = 0$, the potential in the region between the two spheres is given by

$$\Phi(r, \theta, \phi) = \sum_{l=0}^{\infty} \sum_{m=-l}^{l} \frac{a^{l+1}F_{lm}}{a^{2l+1} - b^{2l+1}} \left(r^l - \frac{b^{2l+1}}{r^{l+1}} \right) Y_{l,m}(\theta, \phi).$$

8.23.3 Addition theorem

Given two vectors $\mathbf{r}$, (r, θ, ϕ), and $\mathbf{r}'$, (r', θ', ϕ'), with angle ψ between them,

$$\cos\psi = P_1(\cos\psi) = \cos\theta \cos\theta' + \sin\theta' \sin\theta' \cos(\phi - \phi')$$

and

$$P_n(\cos\psi) = \frac{4\pi}{2n+1} \sum_{m=-n}^{n} Y^*_{n,m}(\theta',\phi')Y_{n,m}(\theta,\phi)\cos(\phi-\phi').$$

8.24 Vector addition (Clebsch–Gordan, Wigner) coefficients

The set of spherical harmonics $Y_{l,m}(\theta,\phi)$ transform according to the irreducible representations $D^{(l)}$ of the rotation group. The eigenvectors of the operator Λ,

$$\Lambda U = \left[\frac{1}{\sin\theta}\frac{\partial}{\partial\theta}\sin\theta\frac{\partial}{\partial\theta} + \frac{1}{\sin^2\theta}\frac{\partial^2}{\partial\phi^2}\right] U = l(l+1)U$$

are represented in Dirac notation by $|l,m\rangle$. The coefficients of the expansion,

$$|J,M\rangle = \sum_{m_1,m_2} (l_1,m_1;l_2,m_2|J,M)|l_1,m_1\rangle|l_2,m_2\rangle,$$

are the vector addition coefficients (Clebsch–Gordan coefficients, Wigner coefficients). [There are several other notations in use: $C(l_1,l_2,J;m_1,m_2,M)$, $(l_1,l_2,J,M|l_1,l_2,m_1,m_2)$, ... , with various permutations, or with punctuation and redundant parameters supressed.]

The coefficients are constrained by the conditions that $2l_i$ and $l_i - m_i$ $(i=1,2)$ are non-negative integers, $0 \le l_i - m_i \le 2l_i$, $M = m_1 + m_2$, and

$$J = |l_1 - l_2|, |l_1 - l_2| + 1, \ldots, l_1 + l_2 - 1, l_1 + l_2$$

so that any of the three values l_1, l_2, J is greater than or equal to the difference between the other two and less than or equal to their sum. This last condition is often expressed by the symbol $\Delta(l_1, l_2, J)$, which is invariant to any permutation of its arguments and is equal to 1 if this "triangle condition" holds and is 0 otherwise.

The coefficients satisfy the symmetry relations

$$\begin{aligned}
(l_1,m_1;l_2,m_2|J,M) &= (l_2,-m_2;l_1,-m_1|J,-M) \\
&= (-1)^{l_1+l_2-J}(l_1,-m_1;l_2,-m_2|J,-M) \\
&= (-1)^{l_1-m_1}\sqrt{\frac{2J+1}{2l_2+1}}(l_1,m_1;J,-M|l_2,-m_2) \\
&= (-1)^{l_2+m_2}\sqrt{\frac{2J+1}{2l_1+1}}(J,-M;l_2,m_2|l_1,-m_1).
\end{aligned}$$

Table 8.3 Clebsch–Gordan (Wigner) coefficients (Ref. 6). In order to conserve space without sacrificing readability, only the matrices for $M \geq 0$ are shown; for $M < 0$, use the symmetry relation

$$(J, -M|l_1, -m_1; l_2, -m_2) = (-1)^{l_1+l_2-J}(J, M|l_1, m_1; l_2, m_2).$$

A $\sqrt{}$ is to be understood over every entry: e.g., 3/5 should be read as $\sqrt{3/5}$, 4/7 should be read as $2/\sqrt{7}$, and $-3/7$ should be read as $-\sqrt{3/7}$. The sign convention is that of Wigner (Ref. 5), Rose (Ref. 7), and Condon and Shortley (Ref. 8).

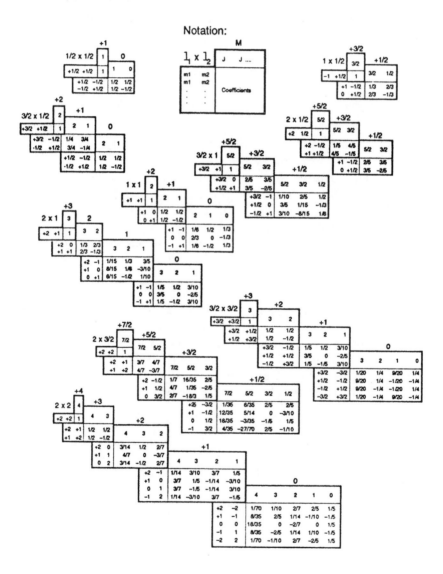

These relations can be expressed more succinctly in terms of the Wigner 3-j symbol[5]

$$(j_1, m_1; j_2, m_2 | j_3, -m_3) = (-)^{j_1 - j_2 - m_3} \sqrt{2j+1} \begin{pmatrix} j_1 & j_2 & j_3 \\ m_1 & m_2 & m_3 \end{pmatrix}$$

with $m_1 + m_2 + m_3 = 0$.

$$\begin{pmatrix} j_1 & j_2 & j_3 \\ m_1 & m_2 & m_3 \end{pmatrix} = (-)^{j_1 + j_2 + j_3} \begin{pmatrix} j_1 & j_2 & j_3 \\ -m_1 & -m_2 & -m_3 \end{pmatrix}$$

$$= (-)^{j_1 + j_2 + j_3} \begin{pmatrix} j_2 & j_1 & j_3 \\ m_2 & m_1 & m_3 \end{pmatrix}$$

$$= (-)^{j_1 + j_2 + j_3} \begin{pmatrix} j_3 & j_2 & j_1 \\ m_3 & m_2 & m_1 \end{pmatrix}.$$

The 3-j symbol is invariant to a cyclic permutation of the columns.

Since the wave functions are orthonormal, the inverse transformation is

$$|l_1, m_1\rangle |l_2, m_2\rangle = \sum_J (l_1, m_1; l_2, m_2 | J, M) |J, M\rangle.$$

Table 8.4 Clebsch–Gordan coefficients for $l_1 \times \frac{1}{2}$.

$$l_1, M - m_2; \tfrac{1}{2}, m_2 | J, M$$

m_2 \ J	$l_1 + \frac{1}{2}$	$l_1 - \frac{1}{2}$
$+\frac{1}{2}$	$\sqrt{\frac{1}{2} + \frac{M}{2l_1+1}}$	$-\sqrt{\frac{1}{2} - \frac{M}{2l_1+1}}$
$-\frac{1}{2}$	$\sqrt{\frac{1}{2} - \frac{M}{2l_1+1}}$	$\sqrt{\frac{1}{2} + \frac{M}{2l_1+1}}$

Table 8.5 Clebsch–Gordan coefficients for l_1.

$$l_1, M - m_2; 1, m_2 | J, M$$

m_2 \ J	$l_1 + 1$	l_1	$l_1 - 1$
$+1$	$\sqrt{\frac{(l_1+M)(l_1+M+1)}{2(l_1+1)(2l_1+1)}}$	$-\sqrt{\frac{(l_1+M)(l_1-M+1)}{2l_1(l_1+1)}}$	$\sqrt{\frac{(l_1-M)(l_1-M+1)}{2l_1(2l_1+1)}}$
0	$\sqrt{\frac{(l_1-M+1)(l_1+M+1)}{(l_1+1)(2l_1+1)}}$	$\frac{M}{\sqrt{l_1(l_1+1)}}$	$-\sqrt{\frac{(l_1-M)(l_1+M)}{l_1(2l_1+1)}}$
-1	$\sqrt{\frac{(l_1-M)(l_1-M+1)}{2(l_1+1)(2l_1+1)}}$	$\sqrt{\frac{(l_1-M)(l_1+M+1)}{2l_1(l_1+1)}}$	$\sqrt{\frac{(l_1+M)(l_1+M+1)}{2l_1(2l_1+1)}}$

The case $l \times \frac{1}{2}$ with $l = 1$ is used to indicate the application of the Clebsch–Gordan coefficients. Thus, reading down the table, the total spin wave functions are expressed in terms of the individual wave functions:

$$\left|\tfrac{3}{2}, -\tfrac{1}{2}\right\rangle = \sqrt{\tfrac{1}{3}} \cdot \left|1, -1\right\rangle\left|\tfrac{1}{2}, \tfrac{1}{2}\right\rangle + \sqrt{\tfrac{2}{3}} \cdot \left|1, 0\right\rangle\left|\tfrac{1}{2}, -\tfrac{1}{2}\right\rangle$$

and, reading across the table, the individual wave functions are decomposed into superpositions of the total spin wave functions:

$$\left|1, 0\right\rangle\left|\tfrac{1}{2}, -\tfrac{1}{2}\right\rangle = \sqrt{\tfrac{2}{3}} \cdot \left|\tfrac{3}{2}, -\tfrac{1}{2}\right\rangle + \sqrt{\tfrac{1}{3}} \cdot \left|\tfrac{1}{2}, -\tfrac{1}{2}\right\rangle.$$

References

[1] *Handbook of Mathematical Functions*, edited by M. Abramowitz and I. Stegun, NBS AMS-55 (U.S. GPO, Washington, D.C., 1964).

[2] I. S. Gradsteyn and I. M. Ryzhik, *Tables of Integrals, Series, and Products* (English translation, A. Jeffrey) (Academic, New York, 1980).

[3] T. W. Körner, *Fourier Analysis* (Cambridge University Press, Cambridge, 1988).

[4] International Organization for Standardization, *ISO Standards Handbook 2, Quantities and Units* (ISO 31/11), Geneva, 1993.

[5] E. P. Wigner (unpublished, 1940) in *Quantum Theory of Angular Momentum*, edited by L. C. Biedenharn and H. van Dam (Academic, New York, 1965); *Group Theory* (Academic, New York, 1959).

[6] E. R. Cohen, thesis, California Institute of Technology, 1949; V. Heine, *Group Theory in Quantum Mechanics* (Pergamon, London, 1960). See also: Particle Data Group, *Review of Particle Properties*, Phys. Rev. D. **50**, 1287 (1994).

[7] M. E. Rose, *Elementary Theory of Angular Momentum* (Wiley, New York, 1957).

[8] E. U. Condon and G. H. Shortley, *The Theory of Atomic Spectra* (Cambridge University Press, New York, 1953).

9

PROBABILITY AND STATISTICS

9.1   Probability distributions

The probability that an "event" occurs is denoted by P; the probability that it does not occur, by $Q = 1 - P$. The probability distribution of a stochastic variable X, $F_X(x)$, is the probability of occurrence of the event "X is less than or equal to x". If X can take a continuous set of values, the *probability density* (or *frequency*) function $f(x)$ is defined such that the probability of the event "$x < X < x + dx$" is $f_X(x)\, dx$.

$$F_X(x) = \int_{-\infty}^{x} f_X(x')\, dx'.$$

If the stochastic variable X takes on only a set of discrete values x_i (finite or infinite), the distribution $F_X(x)$ can be described in terms of delta functions:

$$F_X(x) = \sum_{x_i \leq x} p_i = \int_{-\infty}^{x} \sum_i p_i \delta(x' - x_i)\, dx'.$$

(It is also possible to have a distribution that is the sum of a continuous component and a discrete component.)

Normalization :
$$F(+\infty) = 1 = \sum_i p_i = \sum p(x_i) = \int_{-\infty}^{\infty} f(x)\, dx.$$

Mean :
$$\mu = \sum_i x_i p_i = \sum_i x_i p(x_i) = \int_{-\infty}^{\infty} x f(x)\, dx.$$

Expectation:
$$\mathrm{E}\, g(x) = \langle g(x) \rangle = \mathrm{E}\{g(x)\} = \mu\big(g(x)\big)$$
$$= \sum_i g(x_i) p_i = \int_{-\infty}^{\infty} g(x) f(x)\, dx.$$

183

k-th Moment:
$$m_k = \sum x_i^k p_i = \int_{-\infty}^{\infty} x^k f(x)\, dx$$
(but not all moments may be finite),

k-th Central Moment :
$$\mu_k = \sum (x_i - \mu)^k p_i = \sum (x_i - \mu)^k p(x_i)$$
$$= \int_{-\infty}^{\infty} (x - \mu)^k f(x)\, dx$$

Variance :
$$\sigma^2 = \mu_2 = \sum (x_i - \mu)^2 p_i = \int_{-\infty}^{\infty} (x - \mu)^2 f(x)\, dx,$$

Skewness Coefficient : $\gamma_1 = \mu_3/\sigma^3$,

Excess (Kurtosis): $\gamma_2 = \mu_4/\sigma^4 - 3$.

If $y = g(x)$ $[x = g^{-1}(y) = h(y)]$ is a monotonic function, the distribution $F_Y(y)$ of y is

$$F_Y(y) = \int_{-\infty}^{g^{-1}(y)} f_X(x)\, dx = \int_{-\infty}^{y} f_Y(y)\, dy$$

and the probability density $f_Y(y)$ is

$$f_Y(y) = f_X(h(y))\, h'(y)$$

Confidence Level (CL):   the probability, associated with a parameter δ, that an observed deviation from the mean $x - \mu$ is less than δ, i.e.,

$$CL = \int_{\infty}^{\mu+\delta} f(x)\, dx.$$

For a symmetric distribution, a two-sided confidence level is defined as

$$CL = \int_{\mu-\delta}^{\mu+\delta} f(x)\, dx$$

and the interval $[\mu - \delta, \mu + \delta]$ is called the confidence interval with confidence level CL.

9.2   Convolution of distributions; characteristic function

If X and Y are two independent random variables, the linear combination $Z = X + Y$ is a random variable with a probability density given by the convolution of f_X and f_Y:

$$f_Z(z)\, dz = \int\int f_X(x) f_Y(y) \delta(z - x - y)\, dy dx$$
$$= \int f_X(z - y) f_Y(y)\, dy = \int f_X(x) f_Y(z - x)\, dx \equiv f_X \circ f_Y,$$

$$f \circ g = g \circ f,$$
$$f \circ (g \circ h) = (f \circ g) \circ h = f \circ g \circ h.$$

The operation of convolution can be extended to the case of several functions to give the distribution of the sum $Z = X_1 + X_2 + X_3 + \cdots + X_n$. The Fourier transform of the convolution is the product of the Fourier transforms of its components. The characteristic function of the distribution of X is the expectation of e^{itX} and is essentially the Fourier transform of $f_X(x)$; it is defined by

$$\phi_f(t) = \phi(t; f) = \mathrm{E}\, e^{itX} = \int_{-\infty}^{+\infty} e^{itx} f(x)\, dx.$$

The characteristic function is the generating function for the moments:

$$\phi_f(t) = \phi(t; f) = 1 + \sum_{k=1}^{\infty} m_k \frac{(it)^k}{k!}.$$

9.2.1   Cumulants (semi-invariants)

The logarithm of the characteristic function of a convolution is the sum of the logarithms of the characteristic functions of its components. The *cumulants* K_n of a distribution are defined by the Taylor expansion of the logarithm of the characteristic function:

$$\ln \phi_f(t) = \sum_{k}^{\infty} K_k \frac{(it)^k}{k!}.$$

The first cumulant is the mean, $K_1 = \mu_1 = \mu$; the second cumulant is the variance: $K_2 = \sigma^2$. Except for K_1 the cumulants are invariant to a change of the origin. The kth cumulant of the distribution of the sum of N variables is the sum of the kth cumulants of the component distributions. The standardized cumulants,

$$\gamma_k = \frac{\sum\limits_{j=1}^{N} K_{k+2,j}}{\sigma^{k+2}}, \qquad \sigma^2 = \sum_{j=1}^{N} \sigma_j^2,$$

decrease as $(1/\sqrt{N})^k$ as N increases. This is the essence of the *Central Limit Theorem*: The distribution of the mean of N quantities (drawn from the same or from different distributions) approaches a Gaussian distribution as N increases. More precisely, as the number of significant contributions to a probability distribution increases, the higher cumulants become relatively less important, and the shape of the distribution is dominated by its first two cumulants: the mean $\mu = \sum \mu_j$ and the variance $\sigma^2 = \sum \sigma_j^2$. The distribution approaches a Gaussian shape as $N \to \infty$, since that distribution has only those two non-zero cumulants.

The *skewness*, $\gamma_1 = \mu_3/\sigma^3$, is a measure of the asymmetry of the distribution, and the *excess* or *kurtosis*, $\gamma_2 = \mu_4/\sigma^4 - 3$, is a measure of the spread of the distribution relative to the normal distribution. For $\gamma_2 > 0$ the distribution is higher near the mean and narrower but with longer tails, than the normal distribution, while for $\gamma_2 < 0$ it is flatter and more compact.

9.3   Some common distributions

Much more detailed information on probability distributions including tables, relationships between distributions, and computational algorithms (including Monte Carlo algorithms for pseudo-random variables) is given in Ref. 1, Chapter 26. [For Monte Carlo algorithms, see also Ref. 2, p.III.32.]

9.3.1   Binomial distribution

The binomial distribution $P_n(r)$ is the probability of observing r "successes" out of n independent trials, when the probability of "success" is p (and the probability of "failure" is $q = 1 - p$) for each trial. $P_n(r)$ is the coefficient of t^r in the expansion of $(pt + q)^n$:

$$P_n(r) = \binom{n}{r} p^r q^{n-r} = \frac{n!}{(n-r)!r!} p^r q^{n-r},$$

$$\mathrm{E}\, r = \langle r \rangle = \mu = np,$$

$$\mathrm{E}\, r^2 = n(n-1)p^2 + np, \qquad \sigma^2 = npq,$$

$$\mathrm{E}\left\{r(r-1)(r-2)\cdots(r-k)\right\} = n(n-1)(n-2)\cdots(n-k)p^{k+1},$$

$$\gamma_1 = (q-p)/\sqrt{npq}, \qquad \gamma_2 = (1-6pq)/npq,$$

for $p = q = \frac{1}{2}$: $\gamma_1 = 0,$ $\gamma_2 = -2/n.$

9.3.2   Uniform (rectangular) distribution

If the probability that the random variable lies in a given interval is proportional to the length of the interval within the range $[a, b]$ (with $a < b$), the probability density is

$$f(x) = \begin{cases} 0, & x < a \\ \dfrac{1}{b-a}, & a < x < b \\ 0, & b < x \end{cases},$$

$$\mu = (a+b)/2, \qquad \sigma^2 = (b-a)^2/12, \qquad \gamma_1 = 0, \qquad \gamma_2 = -\tfrac{6}{5}.$$

9.3.3   Normal (Gaussian) distribution

The Gaussian distribution is

$$f(x)\,dx = \frac{1}{\sqrt{2\pi}}e^{-x^2/2}\,dx,$$

$$\mu = 0, \qquad \sigma^2 = 1, \qquad \gamma_k = 0 \quad \text{for all } k \geq 1.$$

More generally, a *normal distribution* with mean μ and variance σ^2 (standard deviation σ) is the distribution of $y = \mu + \sigma x$, where x has a Gaussian distribution, but the two terms are used interchangably.

$$f(y)\,dy = f(y\,|\,m,\sigma)\,dy = \frac{1}{\sqrt{2\pi}\,\sigma}e^{-(y-\mu)^2/2\sigma^2}\,dy.$$

The width of the distribution is also expressed by the "probable error" (symmetric confidence level, CL $= 0.5$), 0.6745σ; mean absolute deviation ($\text{E}\,|x - \mu|$), 0.7979σ; half-width at half-maximum ($\exp[-x^2/2\sigma^2] = \frac{1}{2}$), 1.1774σ.

Confidence levels, and the odds against exceeding a given value of δ/σ are given in the table below:

CL (%)	δ/σ	odds	CL (%)	δ/σ	odds
50	0.6745	$1:1$	95	1.9600	$19:1$
66.667	0.9674	$2:1$	95.45	2.0000	$21:1$
68.27	1.0000	$2.15:1$	99	2.5758	$99:1$
75	1.1503	$3:1$	99.73	3.0000	$369:1$
83.33	1.3830	$5:1$	99.9	3.2905	$999:1$
90	1.6449	$9:1$	99.9535	3.5000	$2148:1$

Confidence levels for the Gaussian distribution may be read from the $\nu = 1$ curve in Figure 9.1 with $\chi^2 = (\delta/\sigma)^2$.

9.3.4   Cauchy distribution

The Cauchy distribution is anomalous in having no finite moments.

$$F(x\,|\,m,a) = \frac{1}{2} + \frac{1}{\pi}\arctan\frac{x-m}{a}, \qquad a > 0,$$

$$f(x\,|\,m,a)\,dx = \frac{a\,dx}{\pi\left[a^2 + (x-m)^2\right]}.$$

This distribution is known as the Lorentz line shape in atomic and molecular physics and the Wigner line shape in nuclear physics.

The generating function for the Cauchy distribution is

$$\phi(t) = e^{imt-a|t|}$$

and the non-differentiability at $t = 0$ shows that the moments,

$$m_k = \lim_{\substack{d \to \infty \\ c \to \infty}} \int_{-c}^{d} x^k f(x \mid m, a) \, dx, \qquad k \geq 1$$

are undefined.

The convolution theorem is still valid (the generating function for a convolution is the product of the generating functions of the components) and this implies that both the centers m and widths a of Cauchy distributions add linearly; an arbitrarily weighted mean of variables drawn from a Cauchy distribution has the same distribution as any individual variable. The sum of N variables drawn from Cauchy distributions has a Cauchy distribution with center $m = \sum m_i$ and width $a = \sum a_i$.

The center m and the width a cannot be estimated from observed moments, but the distribution of the median of N ($N \gg 1$) observations approximates a Gaussian with mean m and variance $\sigma^2 = \pi^2 a^2/4N$, and the width a can be estimated from the semi-quartile range, i.e., one half the range that includes the central half of the data.

9.3.5 χ^2 distribution

The χ^2 distribution is the distribution of the sum of ν components $\sum r_i^2$ in which each r_i is independently normally distributed with unit variance

$$P(\chi^2 \mid \nu) = \frac{1}{2^h \Gamma(h)} \int_0^{\chi^2} t^{h-1} e^{-t/2} \, dt, \qquad h = \nu/2,$$

$$p_\nu(\chi^2) \, d\chi^2 = \frac{(\chi^2)^{h-1}}{2^h \Gamma(h)} e^{-\chi^2/2} \, d\chi^2,$$

$$\mu = \nu, \qquad \sigma^2 = 2\nu,$$

$$\gamma_1 = \sqrt{\frac{8}{\nu}}, \qquad \gamma_2 = \frac{12}{\nu}, \qquad \gamma_k = \frac{(k+1)!}{(\nu/2)^{k/2}}.$$

For $\nu \to \infty$, $x_1 = (\chi^2 - \nu)/\sqrt{2\nu}$ approaches the normal distribution; a better approximation is based on the distribution of $\sqrt{\chi^2}$:

$$x_2 = \sqrt{2\chi^2} - \sqrt{2\nu - 1}$$

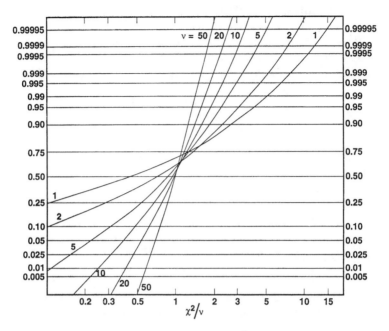

Figure 9.1 Confidence limit CL for the χ^2, normal, and Poisson distributions.

approaches normality more rapidly as $\nu \to \infty$. Still better is the approximation based on the distribution of the cube root of χ^2 given in Ref. 1, Eq. 26.4.14

$$x_3 = \sqrt{\frac{9\nu}{2}} \left(\left[\frac{\chi^2}{\nu} \right]^{1/3} - 1 + \frac{2}{9\nu} \right), \qquad \nu \gg 1.$$

As an example, consider CL = 99.5% (the upper 0.5% tail of the distribution): For $\nu = 50$ the critical value of χ^2 is 79.49. This value of χ^2 gives $x_1 = 2.95$ which corresponds to $1 - \text{CL} = 0.16\%$ for a normal distribution; $x_2 = 2.66$ which corresponds to $1 - \text{CL} = 0.39\%$; and $x_3 = 2.573$ which corresponds to $1 - \text{CL} = 0.503\%$. For $\nu = 5$ the critical value of χ^2 is 16.75. This value of χ^2 gives $x_1 = 3.72$ which corresponds to $1 - \text{CL} = 0.01\%$; $x_2 = 2.79$ which corresponds to $1 - \text{CL} = 0.27\%$; and $x_3 = 2.565$ which corresponds to $1 - \text{CL} = 0.52\%$.

9.3.6 Poisson distribution

The Poisson distribution is a discrete distribution defined by a single parameter, the mean λ. The distribution can be obtained as the limit of the binomial distribution for $p \to 0$ and $n \to \infty$ with $\lambda = np$ held fixed:

$$P_\lambda(r) = \frac{\lambda^r}{r!} e^{-\lambda} \qquad (r = 0, 1, 2, \ldots),$$

$$\mathrm{E}\, r = \mu = \overline{n} = \lambda, \qquad \mathrm{E}\,(r - \lambda)^2 = \sigma^2 = \lambda,$$

$$\mathrm{E}\,\{r(r-1)(r-2)\cdots(r-k)\} = \lambda^{k+1},$$

$$\gamma_1 = 1/\sqrt{\lambda}, \qquad \gamma_2 = 1/\lambda.$$

The Poisson distribution and the χ^2-distribution are related; for integer values of $h = \nu/2$, the χ^2-distribution is

$$P(\chi^2 \,|\, \nu)) = 1 - e^{-\chi^2/2} \sum_{j=0}^{h-1} \frac{(\chi^2/2)^j}{j!} = \sum_{j=h}^{\infty} P_\lambda(j), \qquad \lambda = \chi^2/2.$$

For n even, the probability $\Pr(n \geq n^* | \lambda)$, that n will be equal to or greater than n^* when the mean value is λ, is the CL for the χ^2-distribution corresponding to $\chi^2 = 2\lambda$ and $\nu = 2n^*$. Thus, in Figure 9.1, the ordinate χ^2/ν is also equal to λ/n^* for the Poisson distribution.

9.3.7 Student's distribution

Student's distribution with ν degrees of freedom is the distribution of

$$t = \frac{X}{\sqrt{S^2/\nu}}$$

when X is a normally distributed random variable with unit variance and zero mean and S^2 is an independent random variable following a χ^2 distribution with ν degrees of freedom. The probability that $|\sqrt{\nu}\, X/S|$ will be less than a fixed constant t is

$$A(t | \nu) = \Pr\left\{ \left| \frac{X}{S/\sqrt{\nu}} \right| \leq t \right\} = \frac{\Gamma(h + \frac{1}{2})}{\sqrt{\nu\pi}\,\Gamma(h)} \int_{-t}^{t} \frac{dx}{\left[1 + \dfrac{x^2}{\nu} \right]^{h+1/2}},$$

where $\nu > 0$, $h = \nu/2$.

$$\mu = 0, \qquad \sigma^2 = \frac{\nu}{\nu - 2}, \qquad \gamma_1 = 0, \qquad \gamma_2 = \frac{6}{\nu - 4}.$$

The nth moment exists only for $n < \nu$; for $\nu = 1$, the Student distribution is a Cauchy distribution and has no moments other than the normalization, $m_0 = 1$.

9.4 Statistics, parameter estimation

Modern statistics developed during the early part of this century as a basis for quantifying the variability of general populations, and for inferring, from sampled data, the characteristics of such populations.

9.4.1 Estimate of a mean, known variances

If the set y_n, $n = 1, 2, \ldots, N$, represents values assumed to be drawn from a population with variance σ^2 and a fixed, but unknown, mean μ, the estimate

$$\overline{y} = m_1 = \frac{1}{N} \sum_{n=1}^{N} y_n$$

is the "best" estimate of μ in the sense that, of all linear combinations of the observations y_n with expectation y, it has the smallest variance. The variance of $\overline{y}$ is σ^2/N.

If the observations y_n are drawn from different populations that are known to have the same mean μ, but different variances σ_n^2, an arbitrarily weighted average

$$\overline{y} = \frac{1}{W} \sum_{n=1}^{N} w_n y_n, \qquad W = \sum_{n=1}^{N} w_n$$

is an unbiased estimate of μ. If the weights are chosen to be inversely proportional to the variance of that distribution, $w_n = k/\sigma_n^2$, this estimate has minimum variance, and this minimum variance (maximum statistical weight) is $\sigma^2(\overline{y}) = k/W$.

9.4.2 Joint estimate of mean and variance

If neither the mean nor the variance of the population from which the observations are drawn is known, both must be estimated from the data. An unbiased estimate of σ^2 is provided by

$$s^2 = \frac{1}{N-1} \sum_{n=1}^{N} (y_n - \overline{y})^2 = \frac{1}{N-1} \sum_{n=1}^{N} y_n(y_n - \overline{y})$$

$$= \frac{N}{N-1}(m_2 - m_1^2) \qquad \text{with} \qquad m_k = \frac{1}{N} \sum_{n=1}^{N} y_n^k.$$

The variance of $\bar{y}$ as an estimate of μ is σ^2/N; this is estimated by s^2/N. If y_n are drawn from a normal distribution with variance σ^2, the variance of s^2 is $2\sigma^4/(N-1)$. If the distribution is not Gaussian, there is an additional term, $+\gamma_2\sigma^4/N$, where γ_2 is the kurtosis defined in Section 9.1.

If $N \ggg 1$, a confidence interval for $\bar{y}$ cannot be found using the normal distribution with variance $s^2(\bar{y}) = s^2/N$, since the uncertainty in the estimated variance must be taken into account. The distribution of $\bar{y}/s$ is given by Student's distribution with $\nu = N - 1$ degrees of freedom.

For $\nu \gg 1$, the Student distribution itself can be approximated by the normal distribution of $(y/s)\sqrt{\nu/(\nu + 2)}$.

9.4.3   Estimate of the mean of a finite population

If the population from which items are drawn is finite (samples are drawn from a finite set of discrete items without replacement) the expression for an estimate of the mean of the distribution based on N samples is unchanged, but an estimate of the variance must take the finiteness of the population into account. If P is the size of the population, then

$$\mathrm{E}\, x_i x_j = \mu^2 + \frac{P\delta_{ij} - 1}{P - 1}\sigma^2.$$

The variance of the mean is

$$\mathrm{var}\,(\bar{x}) = \frac{P - N}{(P - 1)N}\sigma^2$$

and the unbiased estimate of σ^2 is

$$s^2 = \frac{P - 1}{(N - 1)P} \sum_{i=1}^{N}(x_i - \bar{x})^2.$$

9.5   Bayesian statistics; inverse probability[3]

The joint probability density of x and y, $p(x,y)$ can be written in two equivalent forms:

$$p(x,y)\,dxdy = p(x|\,y)\,dx\,p(y)\,dy = p(y|\,x)\,dy\,p(x)\,dx$$

where $p(x|\,y)\,dx$ is the conditional probability of the occurrence of x when it is known that y has occurred, and $p(y|\,x)\,dy$ is the probability of the occurrence of y when it is known that x has occurred. When the distribution of x depends on a parameter (or a set of parameters) denoted by

y, the probability density of x for a fixed set of parameters is $p(x|y)$. The unconditioned (or *a priori*) probability densities of x and y are

$$p(x) = \int p(x|y)p(y)\, dy, \qquad p(y) = \int p(y|x)p(x)\, dx$$

and hence

$$p(y|x) = \frac{p(x|y)p(y)}{\int p(x|y')p(y')\, dy'} \qquad \text{(Bayes' Theorem)}.$$

The conditional probability $p(y|x)$ is also known as the *a posteriori* probability, or the prior probability.

Probability theory is concerned with the properties of the distribution of x for a given set of parameters (the mean, variance, etc.). Bayes' theorem was developed in terms of the joint probability distribution of two variables, but it can be reinterpreted to provides the basis for determining estimates of the values of the unknown parameters from a set of observations, $p(y|x)$ when the observed data are considered to be given quantities rather than sampled random variables. It thus provides the basis for the solution of the inverse of the direct problem of traditional probability.

9.6   Parameter estimation; least-squares fitting

In many situations the data to be analyzed consist of observations y_n ($n = 1, 2, \ldots, N$) of expressions $g_n(x_1, x_2, \ldots ; p_{j,n})$ containing M unknown other quantities (parameters) x_i. The functions g_n may be the same function evaluated for different values of the known parameters p_j, or the set $g_n(x; p)$ may contain physically different functions. For the application of least squares it is only necessary that at least one subset of M expressions $y_n = g_n(x_i; p_{j,n})$ can be solved for a unique set of values of the M unknowns x_i. Then, if $N > M$, the least-squares algorithm will provide the best estimate of the set of M unknown quantities in the same sense that an appropriately weighted mean provides the best estimate of the value of a single variable.

9.6.1   Linear least-squares fit

The simplest situation involves expressions in which the quantities to be determined appear linearly. A typical example is the determination of the coefficients a_i in the expansion of an expression in terms of a set of base functions: $y(z) \approx Y(z) = \sum a_i f_i(z)$. The given data are the pairs (z_n, y_n), $n = 1, 2, \ldots, N$ and the weights w_n. If the observations are independent (two different observations y_n and y_m are uncorrelated) $w_n =$

$1/\sigma_n^2$, the reciprocal of the variance of the distribution from which the observation y_n is assumed to be drawn, under the assumption that the value of z_n is exact. The weights may also be relative values ($w_n \sim 1/\sigma_n^2$) whose normalization, s_o^2 (the variance to be assigned to unit weight), will be determined from the "goodness of fit" of the data. The estimate of the variance σ_n^2 is then given by $s_n^2 = s_o^2/w_n$.

The fitted coefficients a_i are found from

$$a_i = \sum_j V_{ij} Z_j, \qquad Z_j = \sum_n w_n f_j(z_n) y_n,$$

where V_{ij} are elements of the covariance matrix V, the inverse of the weight matrix W:

$$(V^{-1})_{ij} = W_{ij} = \sum_n w_n f_i(z_n) f_j(z_n).$$

The variance of a fitted value of $y(z)$ (interpolated or extrapolated) is given by

$$\sigma^2(y) = \sum_{i,j} f_i(z) V_{ij} f_j(z).$$

It is convenient to use an orthonormal basis for the set $f_i(z)$ but the standard orthogonal functions (such as Legendre polynomials) are defined over a continuous interval, and the orthonormality expressed by

$$\int f_i(z) f_j(z) \, dz = \delta_{ij}$$

will be only approximately valid for the finite sum. The matrix W is therefore not a diagonal matrix even for $w_n = $ constant; but if the data points z_n are appropriately selected, orthogonality can be maintained. Alternatively, polynomials orthogonal over the specific data set can be generated. If the basis set is orthogonal over the data, the weight matrix W is diagonal and the expansion coefficients are independently evaluated: $a_i = Z_i/W_{ii}$. In general, with an approximately orthogonal basis, the weight matrix W will be approximately diagonal, and the off-diagonal terms will be small ($W_{ij}^2 \ll W_{ii} W_{jj}$). This has advantages for computational precision ($V_{ii} \approx (W_{ii})^{-1}$) as well as advantages in the physical interpretation of the fitted coefficients, which are then approximately statistically independent.

The "goodness of fit" of the data is measured by S^2:

$$S^2 = \sum_{n=1}^{N} w_n (y_n - Y_n)^2 = \sum_{n=1}^{N} y_n w_n (y_n - Y_n)$$

$$= \sum_{n=1}^{N} w_n y_n^2 - \sum_{i=1}^{M} a_i Z_i$$

$$= \sum_{n=1}^{N} w_n y_n^2 - \sum_{i=1}^{M} a_i W_{ij} a_j = \sum_{n=1}^{N} w_n y_n^2 - \sum_{ij}^{M} Z_i V_{ij} Z_j.$$

If the weights w_n are the reciprocals of the variances of the input data y_n, the distribution of S^2, considered as a sampling from a population defined by the distributions of y_n, is the χ^2-distribution with $\nu = N - M$ degrees of freedom. If $(S^2 - \nu)\sqrt{2\nu} \gg 1$ the fitted residuals are inconsistent with the precision of the observations and the fit can be improved by including additional functions f_i in the basis set. Basis functions should be retained or added to the fit if their inclusion produces a significant decrease in the value of S^2/ν. However, if the basis set is not orthogonal, the presence of an addition basis function will alter the contributions of the existing functions to the value of S^2.

If w_n are only relative weights, S^2 is only proportional to the variable of a χ^2 distribution and can not be used as a basis for evaluating "goodness-of-fit." The weights (and the variances) can be normalized by multiplying each w_n by the factor ν/S^2, and the covariance matrix V by the factor S^2/ν.

a. Fitting Data to a Straight Line:

If the data set (x_n, y_n, w_n), $n = 1, 2, \ldots, N$ is to be fitted to a straight line, $y = a + bx$, the fitted values of a and b are

$$a^* = \frac{1}{D}[S_y S_{xx} - S_{xy} S_x], \qquad b^* = \frac{1}{D}[S_1 S_{xy} - S_x S_y]$$

with

$$S_1 = \sum_n w_n, \qquad S_x = \sum_n w_n x_n, \qquad S_{xx} = \sum_n w_n x_n^2,$$

$$S_y = \sum_n w_n y_n, \qquad S_{xy} = \sum_n w_n x_n y_n, \qquad S_{yy} = \sum_n w_n y_n^2,$$

$$D = \begin{vmatrix} S_1 & S_x \\ S_x & S_{xx} \end{vmatrix} = S_1 S_{xx} - S_x^2.$$

The covariance matrix of the fitted coefficients is

$$\begin{pmatrix} V_{aa} & V_{ab} \\ V_{ba} & V_{bb} \end{pmatrix} = \begin{pmatrix} s^2(a) & s_{ab} \\ s_{ba} & s^2(b) \end{pmatrix} = \frac{1}{D} \begin{pmatrix} S_{xx} & -S_x \\ -S_x & S_1 \end{pmatrix}.$$

The goodness of fit is

$$S^2 = \sum_n w_n(y_n - a^* - b^* x_n)^2 = S_{yy} - a^* S_y - b^* S_{xy}$$

$$= \frac{1}{D} \begin{vmatrix} S_1 & S_x & S_y \\ S_x & S_{xx} & S_{xy} \\ S_y & S_{yx} & S_{yy} \end{vmatrix}$$

with $\nu = N - 2$ degrees of freedom.

The variance $s^2(y)$ of a point on the fitted line $y = a^* + b^* x$ is given by

$$s^2(y) = \left[\frac{S^2}{\nu}\right] \frac{S_{xx} - 2S_x x + S_1 x^2}{D} = \left[\frac{S^2}{\nu}\right] \left(\frac{1}{S_1} + \frac{S_1(x - \bar{x})^2}{D}\right),$$

where $\bar{x} = S_x/S_1 = \sum w_n x_n / \sum w_n$ is the mean value of x in the data. If the weights w_n are the reciprocals of the variances of the observed values y_n, the term in square brackets may be omitted and the variance evaluated on the basis of "internal consistency"; including the first term produces the variances evaluated on the basis of "external consistency." If a straight line is a valid representation of the data, and if the data are in fact independent with variances $1/w_n$, the two expressions are equally valid estimates of the variance since in this case the expectation of $S^2 = \chi^2$ is $\nu = N - 2$.

If χ^2/ν lies outside the acceptable limits, the data should be examined for possible inconsistencies or refitted to some other curve, or the weights may be accepted as relative values. When the assigned weights are not based on *a priori* information on the variances of the data but have only relative significance, the term S^2/ν (which is an evaluation of the uncertainty to be associated with unit weight) normalizes the variances of the data based on the mean square deviations of the data from the fitted straight line.

b. Non-Linear Data, Linear Approximation:

When the functional relations between the observations and the unknown quantities are not linear, the least-squares algorithm may be applied to a linearized set of equations. Let x_i° be a set of approximations to the unknown quantities x_i, and introduce a new set of quantities u_i and z_n. The simplest choice is $x_i = x_i^\circ + u_i$ and $y_n = g_n(x_i^\circ) + z_n$. A first-order Taylor expansion of the observational equations then leads to a linearized set

$$z_n = \sum_{i=1}^{M} \frac{\partial g_n}{\partial x_i}\Big|_{x_j^\circ} u_i + e_n,$$

where e_n is the residual error.

Equivalent expressions are obtained by writing either $x_i = x_i^\circ(1 + u_i')$ and $y_n = g_n(x_i^\circ)(1 + z_n')$ or $u_i'' = \ln(x_i/x_i^\circ)$ and $z_n = \ln(y_n/g_n(x_i^\circ))$.

The solution of this set of equations has the same form as the solution given above for linear least squares with the identification

$$W_{ij} = \sum_{n=1}^{N} \frac{\partial g_n}{\partial x_i} w_n \frac{\partial g_n}{\partial x_j}, \qquad Z_j = \sum_{n=1}^{N} \frac{\partial g_n}{\partial x_i} w_n z_n,$$

and

$$u_i = \sum_{i=1}^{M} V_{ij} Z_j$$

where $V = W^{-1}$.

c. Matrix Formulation:

Let the observational equations (or their linearized approximation) be expressed as

$$z = y - y^\circ = Au + e,$$

where y is a vector of observed data, $x^\circ + u$ is the vector of parameters to be fitted in the least-squares adjustment, and A is a known matrix. The elements of A are $\partial y_n / \partial x_i$, an $N \times M$ matrix evaluated at the set of fixed values from which the data z_n and u_i represent deviations. The *input covariance matrix* v comprises the variances and covariances of e for the input data ($N \times N$ matrix). The input weight matrix is the reciprocal matrix $w = v^{-1}$. The output estimates of the least-squares fit are

$$u^* = VA^T wz,$$

where $\mathbf{V} = (\mathbf{A}^T \mathbf{w} \mathbf{A})^{-1}$ and T indicates the transposed matrix. The weight matrix for the fitted quantities is $W = A^T w A$ and the covariance matrix is $V = W^{-1}$. The adjusted data are

$$y^* = y^\circ + Au^* = y^\circ + AVA^T w(y - y^\circ) = y^\circ + P(y - y^\circ).$$

The data $z = y - y^\circ$ span a vector space with metric w, in which space length can be defined:

$$|a| = \sqrt{a^T w a}$$

and two vectors a and b are "orthogonal" if $a^T w b = 0$. The matrix $P = AVA^T w$ is a projection operator:

$$PP = P \quad \text{and clearly,} \quad wP = P^T w.$$

The residuals of the fitting are

$$r = y - y^* = z - z^* = z - Au^*.$$

The space of the data is separated into two orthogonal subspaces, $\boldsymbol{Pz}$ and $\boldsymbol{r} = \boldsymbol{z} - \boldsymbol{Pz}$:

$$(\boldsymbol{Pz})^T \boldsymbol{wr} = \boldsymbol{z}^T \boldsymbol{P}^T \boldsymbol{w}(\boldsymbol{z} - \boldsymbol{Pz}) = \boldsymbol{z}^T(\boldsymbol{wP} - \boldsymbol{wPP})\boldsymbol{z} = 0.$$

The goodness-of-fit is given by

$$S^2 = \boldsymbol{r}^T \boldsymbol{wr} = \boldsymbol{z}^T \boldsymbol{wr} = \boldsymbol{z}^T \boldsymbol{wz} - \boldsymbol{u}^{*T} \boldsymbol{Wu}^*.$$

9.7   Error propagation

Errors in the variables of a function will produce errors in its evaluation. If the set x_i are random variables, the quantity $y = f(x_1, x_2, \ldots)$ will be a random variable.

Let the error in x_i be ϵ_i; then η, the error in y, is

$$\eta = \sum_i \frac{\partial f}{\partial x_i} \epsilon_i + \frac{1}{2} \sum_{i,j} \frac{\partial^2 f}{\partial x_i \partial x_j} \epsilon_i \epsilon_j + \cdots .$$

If the errors ϵ_i have means δ_i (systematic component) and a covariance matrix (random component)

$$v_{ij} = \overline{(\epsilon_i - \delta_i)(\epsilon_j - \delta_j)} = \overline{\epsilon_i \epsilon_j} - \delta_i \delta_j,$$

the mean value and variance of y are

$$\bar{y} = f(x_i^\circ) + \sum_i \frac{\partial f}{\partial x_i} \delta_i + \frac{1}{2} \sum_{i,j} \frac{\partial^2 f}{\partial x_i \partial x_j}(\delta_i \delta_j + v_{ij}) + \cdots$$

$$= f(x_i^\circ + \delta_i) + \frac{1}{2} \sum_{i,j} \frac{\partial^2 f}{\partial x_i \partial x_j} v_{ij} + \cdots$$

and

$$\sigma^2(y) = \sum_{i,j} \frac{\partial f}{\partial x_i} \frac{\partial f}{\partial x_j} v_{ij} + \cdots .$$

If the random components of the errors are uncorrelated, $v_{ij} = \sigma_i^2 \delta_{ij}$, the variance becomes

$$\sigma^2(y) = \sum_{i,j} \left(\frac{\partial f}{\partial x_i}\right)^2 \sigma_i^2 + \cdots .$$

For a function of a single variable x with variance σ^2, the mean and variance of y are

$$\bar{y} = f(x) + \frac{1}{2} f''(x)\sigma^2 + \cdots , \qquad \sigma^2(y) = f'(x)^2 \sigma^2.$$

Since, in the evaluation of experimental data, the true means x_i are not known but are only approximated, the second derivative terms in $\bar{y}$ may often be omitted since they can be much smaller than the unknown contributions from the first-order errors in $\bar{y}$.

References

[1] *Handbook of Mathematical Functions*, edited by M. Abramowitz and I. A. Stegun (U.S. GPO, Washington, D.C., 1967).

[2] Particle Data Group, Review of Particle Properties, Phys. Rev. D **50**, 1173 (1994).

[3] H. Cramér, *Mathematical Methods of Statistics* (Almquist and Wiksells, Uppsala, 1945; Princeton University Press, Princeton, 1946).

[4] *Guide to the Expression of Uncertainty in Measurement* (International Organization for Standardization, Geneva, 1993).

[5] M. G. Kendall and A. Stuart, *The Advanced Theory of Statistics, Distribution Theory* Vol. 1, 3rd ed., 1969; Vol. 2, *Inference and Relationship*, 2nd ed., 1967; Vol. 3, *Design and Analysis, and Time Series*, 2nd ed., 1968. (Griffin and Company, London).

SUBJECT INDEX